Hermaphrodites and the
Medical Invention of Sex

Hermaphrodites
and the
Medical Invention
of Sex

Alice Domurat Dreger

HARVARD UNIVERSITY PRESS
Cambridge, Massachusetts
London, England
1998

Library of Congress Cataloging-in-Publication Data
Dreger, Alice Domurat.
Hermaphrodites and the medical invention of sex / Alice Domurat Dreger.
p. cm.
Includes bibliographical references and index.
ISBN 0-674-08927-8 (alk. paper)
1. Hermaphroditism—Treatment—France—History—19th century.
2. Hermaphroditism—Treatment—Great Britain—History—19th century.
3. Hermaphroditism—Treatment—France—History—20th century.
4. Hermaphroditism—Treatment—Great Britain—History—20th century.
5. Hermaphroditism—Treatment—United States.
6. Hermaphroditism—Psychological aspects. 7. Sex (Psychology) I. Title.
RC883.D695 1998
616.6'94—DC21   97-40487

CONTENTS

ILLUSTRATIONS

ACKNOWLEDGMENTS

IN ONE OF MY FAVORITE MOVIES, *Harvey,* the main character, Elwood P. Dowd (played by Jimmy Stewart), says that his mother always
used to tell him that "in this world you must either be Oh So Smart, or
Oh So Pleasant." Poor Mrs. Dowd obviously never encountered the sort
of people I have been lucky enough to meet during the seven years this
work has been in progress. Day after day, this book has allowed me to
interact with people of every background who are both wise and kind.

The topic was suggested to me early in my graduate career by Fred
Churchill. Fred Churchill and Ann Carmichael served as co-chairs of my
dissertation committee, and I am extremely grateful to them and to the
two other members of my committee, Jeanne Peterson and Richard Sor-
renson, for their guidance, wisdom, encouragement, and generosity. I feel
privileged to have earned my Ph.D. in 1995 from the Department of
History and Philosophy of Science at Indiana University, where I received
from my classmates and professors a diverse and lively education. Stephen
Kellert helped me to think more clearly about gender, reality, and responsibility; Jim Capshew encouraged me to explore the often-neglected question of how science has shaped culture; and Noretta Koertge challenged
me to think about the importance of truth and honesty. My fellow graduate students gave me the kind of comfort only peers can. I am especially
grateful to Carol Engelhardt, Elizabeth Green, Vernon Rosario, and Judy
Johns Schloegel for references, ideas, and company. Theresa Anderson
and Karen Blaisdell also provided me with support, not least of which
included hours of assistance with administrative and funding issues.

The Program in History of Science and Technology at the University of Minnesota granted me a visiting professorship in 1995–96, and that position proved invaluable to the production of this book. I am grateful to the students, faculty, and staff of the program, particularly John Eyler, Sally Gregory Kohlstedt, and Alan Shapiro, and the members of my graduate biomedical ethics reading group there—Andrew Burnett, Michelle Hirschboeck, David Pedersen, and Effy Vayena.

For their generous support and assistance during the latter stages of the project, I am grateful to George and Ellen Leroi and the students, staff (especially Sandy Conner, Ed Ingraham, and Gerry Smith), and faculty (especially Robert Shelton) of the Lyman Briggs School in the College of Natural Science of Michigan State University. Jessica Berg, Libby Bogdan-Lovis, Howard Brody, Janet Burstein, and other members of the MSU Center for Ethics and Humanities in the Life Sciences have also been extremely kind and helpful.

This project could not have been started, or completed, without the generous help of a large number of curators and librarians. At the Wangensteen Biomedical Historical Library of the University of Minnesota, the support of curator Elaine Challacombe and her assistant Maria Falconer was unending and outstanding. My thanks go also to Dr. Stephen Greenberg of the National Library of Medicine's History of Medicine Division, Barbie Wells of the Anatomy Library at Cambridge University, Nancy Eckerman of the Ruth Lilly Medical Library at Indianapolis, and Gretchen Worden of the Mütter Museum at the College of Physicians of Philadelphia. I am indebted as well to the staffs of the British Library, the Bibliothèque Nationale in Paris, the Cambridge University Library, the National Library of Medicine, Interlibrary Loan Services, and the libraries of Indiana University, the University of Minnesota, and Michigan State University.

A few of the findings and analyses presented in this book were included in "Doubtful Sex: The Fate of the Hermaphrodite in Victorian Medicine," *Victorian Studies,* 38, no. 3 (Spring 1995): 335–370), and are reprinted here by permission of the Trustees of Indiana University. Chapter 4 is a lengthened adaptation of "Hermaphrodites in Love: The Truth of the Gonads," in *Science and Homosexualities,* ed. Vernon A. Rosario (New York: Routledge, 1997); I am grateful to Vernon for his assistance with that essay.

Funding for this project has been generously provided by Indiana University (including a John Edwards Fellowship and a travel grant from the International Programs Office), the University of Minnesota Program in History of Science and Technology, Michigan State University, and my parents (who may not understand what I do, but who love and believe in me anyway). A critical year of work was funded by a Charlotte Newcombe Doctoral Dissertation Fellowship from the Woodrow Wilson Foundation.

Additional people gave me important material, advice, and/or pre-prints of their work. I am indebted to Diane Marie Anger, Cheryl Chase, Martha Coventry, Anne Fausto-Sterling, Morgan Holmes, Stephanie Kenen, Suzanne Kessler, Justine Schober, and others too numerous to name. Key feedback came from Howard Brody, Daniel Federman, Bo Laurent, Vernon Rosario, Londa Schiebinger, Justine Schober, and an anonymous reviewer for Harvard University Press. My editor at Harvard University Press, Michael Fisher, has proved patient, wise, and kind. Elizabeth Gretz, manuscript editor, has served as this book's faithful midwife and godmother. Ann Downer-Hazell also has provided important editorial assistance.

I regret that it is impossible to name here all of the people who have given me moral and intellectual support for this project. With all the help I have been given, the usual should go without saying, but I will say it anyway: any remaining omissions and errors are my own responsibility.

Finally, I dedicate this work to my husband, Aron Sousa, who never let medical school or his residency get in the way of providing me with love, ideas, advice, support, a wonderful second family, and the occasional batch of clean laundry. Throughout the last four years, Aron has fed my body, mind, and soul. He is truly Hermaphroditos to my Salmacis, and Salmacis to my Hermaphroditos, and I am thrilled that the gods chose to stick us together.

Hermaphrodites and the
Medical Invention of Sex

PROLOGUE

"But My Good Woman, You Are a Man!"

THE TITLE OF THIS PROLOGUE comes from the remarkable encounter of two people at a Belgian surgical clinic one winter day roughly eleven decades ago. The encounter was reported in an 1886 medical review journal under the provocative title, "An Example of Error of Sex Owing to Apparent Hermaphrodism." The report's author, a medical doctor by the name of Dandois, related the recent meeting between a colleague of his, Professor Michaux, and the alleged subject of mistaken sex, a seeming woman identified in the record only as Sophie V.

It happened that on the ninth of February immediately preceding the published report, Sophie, a domestic servant who was then forty-two, went to the local surgical clinic seeking advice and help. She had now been married two months to her first husband and, in spite of the couple's best efforts, Sophie's husband could not "accomplish the conjugal act with her" to his satisfaction; he could not seem to penetrate her vagina.[1] Sophie wanted to know what was wrong, and she wanted to know if the problem could be fixed.

Professor Michaux, on duty at the clinic that day, examined Sophie's genitals and quickly came upon what he thought was the source of the problem: Sophie V. was really a man, no matter what she had been led to believe all of her life, no matter what her husband had been led to believe. Sophie had, according to Michaux, a penis about five centimeters long "existing in the usual place," and, although it was lacking a hole at the tip, it was, like most penises, capped with a glans and it was, as the published report delicately put it, "susceptible to erection." Sophie's urine apparently exited via a hole near the base of her "penis." Meanwhile, her labia—what Michaux decided were really a divided scrotum—contained at least one testicle, in Michaux's opinion. After all, this supposed testicle felt and behaved quite like the testicles in Michaux's previous experience. And although Sophie did not have the dense facial hair that usually predominated in a man her age, neither did she have all the expected signs of femininity; Sophie herself confessed that her husband had "breasts much greater than she." Finally, Michaux noted, of course Sophie's husband couldn't get his penis into her vagina, for she didn't have one.[2]

Again and again, Michaux insisted to the incredulous Sophie that the problem was not her anatomy but her understanding of her anatomy. He insisted that he knew the truth of her flesh and bones and that she ought to face facts. Sophie could not believe what she was hearing. Still, Michaux repeated to his stunned patient, "But my good woman, you are a man!"[3]

We do not have a first-hand account of Sophie's impressions of this meeting, but the story as told by Dandois hints that Sophie thought Michaux was either strangely cruel or a bit daft. Even Dandois conceded:

> It is difficult to figure what should be the sentiments of a person [like Sophie], engaged in the bans of marriage, to whom it is declared point-blank that she herself, her spouse, her near relations, her friends, her entourage, the civil and religious authorities, are [all] mistaken about her sex, and that she is the victim of an error of which for forty years everyone has partaken, and which even she herself could not suspect![4]

True, there had been some question about Sophie's sex when she was born, and her parents had taken her to a medical man when she was a few weeks old. Apparently at that time "the man of the art did not find the thing sufficiently clear, and he asked the parents to return later with the child."[5] Sophie's parents did not, though, because they feared the man might accidentally hurt or even kill her in his investigations. Instead they decided Sophie must be a girl—her genitals looked fairly feminine to them at the time—and so she was raised a girl. When she was in her early twenties, Sophie had developed what she thought was a hernia, and she wore supportive bandages from that time forward. But now, according to Michaux, that "hernia" was really a descended testicle. And now she was married to a man. What to do?

To Michaux, the answer seemed simple. Sophie had testicles and a penis; she was surely a man; she was therefore not really married (no matter what she and her husband and their friends and families thought), because no marriage between two men was a true or legal marriage. She—or rather he—should have her civil status formally changed to male, her marriage officially annulled, and should start acting her "true" sex. Sophie was a man.

To Sophie, the answer was apparently equally simple. She felt like a woman, she dressed like a woman, and she had always been a woman. Even if some of her anatomy looked a little unusual, so what? What business was that of others? She was a woman. She was married to a man whom she loved as a husband and who loved her as a wife. The medical man to whom she now had come for help must be either crazy, wrong, confused, or at the very least not worth heeding, because Sophie had no interest in suddenly becoming a man.

In Michaux's mind, of course, the issue was not one of *becoming* a man. Sophie's anatomy—from the doctor's perspective, most especially her testicles—indicated that she was already a man and had always been one. This frustrating "good woman" simply refused to accept the fleshy truth.

HUMANS COME IN a wonderful array of types: many sizes, many abilities, many features, and many approaches to experiencing and organizing the world. Human variation is truly extraordinary. Amid all that variation, we can say with some confidence that most human beings develop one of two common sets of a particular group of organs. That is, most people possess either what is now labeled "female" or "male" sexual anatomy. What to count as the key sexual anatomy can be problematic, as we shall see, but we can generally agree that certain physiologically and socially important organs are usually found only in either those called males or those called females. Most of the people designated female are born with a clitoris, ovaries, fallopian tubes, a vagina leading to a uterus, a separate outlet for the passage of urine, labia minora and majora, and so on. We also have known, since earlier this century, that most females have what is called an "XX" chromosomal (genetic) pattern.[6] Meanwhile most of the people designated male are born with testicles contained in a scrotal sac, vas deferens, a prostate, a penis through which the urethra passes to open at the tip of the glans, and so on, and most of them have what is called an "XY" chromosomal pattern. Yet some people—more than is generally assumed—are born with an anatomical conformation different from "standard" male or female bodies. Their unusual anatomies can result in confusion and disagreement about whether they should be considered female or male or something else. These people have for centuries been labeled "hermaphrodites," and since the early 1900s they have also sometimes received the medical designation of "intersexual."

Many people who are unfamiliar with intersexuality assume that the most striking variations in sexual anatomy must be due to some associated variation in the usual XX or XY chromosomal patterns. They often assume that intersexuals must have an extra or a missing "sex" chromosome. We live in an age of genetics and oversimplified stereotypes about the nature of males and females, so it is not surprising that many people assume there must be a simple genetic, algebraic sort of solution to sex variation.[7] But all sexuality is complicated, and intersexuality is no exception. Although it is true that a very small percentage of the people not easily sorted into male or female have been shown to have chromosomal patterns that differ from the common XX and XY varieties, the majority do appear, upon examination of their chromosomes, to have the standard male or female "sex chromosome" pattern. (I place "sex chromosome" in quotation marks because the term is an unfortunate misnomer leading to much confusion among people unfamiliar with genetics; genes related to traits we consider nonsexual are also located on X chromosomes, and genes located on chromosomes besides the X and Y chromosomes contribute to sexual development. We would do better to call these X and Y chromosomes instead of "sex chromosomes.") Today experts believe that intersexuality can be caused by a wide range of conditions, discussed in Chapter 1. The sexual development of any individual is a complicated and amazing event, involving the working of chromosomes, the action of self-produced and/or ingested hormones, the effects of environmental agents like toxins and nutritional substances, social norms like those that dictate circumcision or clitoridectomy (removal of the clitoris) or certain levels of physical prowess, family dynamics, individual sexual encounters, accidents, and so on. What it means to be a male, a female, or a hermaphrodite—or what it means to become a male, a female, or a hermaphrodite—goes far beyond the "sex chromosomes."

Indeed, when we focus on hermaphrodites, as this book does, we sometimes forget how much variation in sexual anatomy there is among undoubted males and females. Clitorises and penises, for instance, come in a wide range of shapes and sizes even in people labeled "normal" in terms of their sex. Some phalluses look different because they have been subjected to disease, enhancement, or alteration. Some simply grow to be relatively small or large, colorful or monochrome, thick or thin. The standardized pictures found in anatomy textbooks and on gynecologists' walls make it seem as if all female genitals look pretty much alike and all male genitals pretty much alike, but in fact any gynecologist, urologist, or pediatrician can tell you that in actual practice things are not so cut and dried. Genitals vary a great deal.[8] And there is quite a bit of variation in what we think of as "sexual" anatomy beyond the genitals, too. For instance, breasts are often thought of as a "feminine" trait, yet a good number of women are quite flat-chested, and big-breastedness is common enough in men in the United States today that it has warranted a whole sequence of jokes on the television comedy "Seinfeld," in an episode in which two characters try to design and market a bra for men (to be called either "The Bro" or "The Mansierre"). Similarly, facial hair is thought of as a masculine trait, yet there are scores of products available in every drugstore marketed specifically to women for the removal of facial hair.

When we look at hermaphrodites, we are forced to realize how variable even "normal" sexual traits are. Indeed, we start to wonder how and why we label some traits and some people male, female, or hermaphroditic. We see that boundaries are drawn for many reasons, and could be—and have been—drawn in many different ways, and that those boundaries have as many complex effects as they do causes.

SOMETIMES WHEN PEOPLE ask how I got interested in the history of hermaphrodites I tell them a tale from my own life history that occurred roughly ninety years—one long lifetime—after the encounter of Sophie and Michaux. There I was at the age of thirteen or so, sitting in my eighth-grade health class, squirming nervously with the rest of my classmates. We were starting the unit on sex education, and we were all scared stiff of what the (by all accounts) completely uninhibited teacher was going to make us discuss. For a moment he said nothing, only adding to the tension. Finally he spoke.

"How do you know if you're a boy or a girl?" We cringed and giggled. Surely he was not going to make us say it. He rephrased his question slightly: "How can you tell if you're a boy or a girl?"

Awkward pauses and sideways glances. Eventually one student, the nerdy fellow in the front of the class, broke the tense silence: "By the genes!" We all sighed in relief at this lovely, safe, boring answer.

"Right!" the teacher responded without a moment's hesitation. "You pull 'em down! You pull those jeans down!"

We burst out laughing, our faces all bright red.

After that joke, long after I stopped laughing and for a long time before I learned about hermaphrodites, the question bothered me. What, exactly, made me a girl? My mother often told me I could do or be anything I wanted, yet I knew very well that even day-to-day things depended on my sex: which bathroom I would use, whether I could wear dresses or be seen crying, what my friends or my family would say if I linked arms with a girl or with a boy. I was not even sure whether my mother was right that I could be or do anything in spite of being a girl. The recent defeat of the Equal Rights Amendment baffled me. If my sex really didn't matter in terms of my rights, as Mom insisted, why were my compatriots not willing to say so? Did it matter that I was a girl? How did it matter? And what, exactly, did it mean to be a girl? How would this body shape my life? How would changes in my body change my life—and *why* would changes in my body change my life?

Most of us are used to the idea of two sexes, and most of us are used to being only one, even if, as we all know, some people refuse or fail to fit into the generally accepted behavior pattern for what most people think of as "normal" masculinity or femininity. But the fact is that not everybody arrives in this world ready to be squeezed into one or the other generally accepted *anatomic* patterns of what we usually think of as male and female. Hermaphroditism causes a great deal of confusion, more than one might at first appreciate, because—as we will see again and again—the discovery of a "hermaphroditic" body raises doubts not just about the particular body in question, but about all bodies. The questioned body forces us to ask what exactly it is—if anything—that makes the rest of us unquestionable. It forces the not-so-easy question of what it means to be a "normal" male or a "normal" female. Georges Canguilhem noted in his study of the normal and the pathological, "it is not paradoxical to say that the abnormal, while logically second, is existentially first."[9] In other words, we tend to assume that the normal (in this case the "normal" sexual anatomy) existed before we encountered the abnormal, but it is really only when we are faced with something that we think is "abnormal" that we find ourselves struggling to articulate what "normal" is. Although it seems not so difficult to recognize that some bodies look fairly unusual in terms of their genitalia, it is difficult to answer the question of what exactly a hermaphrodite is, because to do so one must first decide what trait or traits are so important to femalehood and malehood that the possession of a combination of those traits by any single body would necessarily designate that single body hermaphroditic.

What, then, are the necessary or sufficient traits of femalehood or malehood? Don't ask the International Olympic Committee. They check "women" athletes to make sure there are no men masquerading as women, but the committee's decisions about how to tell who is a "real" woman keep changing and keep getting challenged.[10] At first modern Olympic officials let athletes divide themselves into males and females, but some (like Hermann Ratjen in 1936, who lost anyway) seemed to be cheating. The Olympic organizers then started to use genital exams to sort male from female athletes, but not everyone was easy to sort by that means; more than a few cases seemed too confusing to let a genital exam suffice. In 1968, when it became feasible, Olympic officials moved to buccal smears and a look at the "sex chromosomes" of would-be women competitors. It seemed that surely this procedure would identify each athlete's "true" sex. But not necessarily. Some people, for example, have an XY (so-called male) chromosomal basis but develop more along "feminine" pathways because their bodies lack the physiological receptors to read and respond to the testosterone their bodies produce. These people are called "androgen insensitive." Should we count them as men, even though they have the look and experience of being women, even though the physiological effects of their hormones are not like those of the common male athlete? If anything they should be given extra credit by this logic, since the bodies of XX women do produce and respond to testosterone, thought of as a "strength-building" hormone, while the bodies of these XY women respond to it incompletely or not at all. Measure the hormones (or the response to them), then, and gauge everyone accordingly? But the "sex" hormones don't divide simply into two kinds, "male" and "female." Men and women produce the same kinds of hormones, though usually in different relative quantities, but we know that all girls' and women's bodies do not uniformly produce a single, identifiable "feminine" cocktail of hormones, nor do all boys and men produce a single, identifiable sort of "masculine" cocktail.[11] To use hormones as a dividing line we would have to decide where to draw the boundaries of acceptable hormonal variations on malehood and femalehood. That too would be tricky, and perhaps particularly difficult to justify. And again, what if a person produced certain hormones to which her body did not react? What to do then? Measure production *and* reaction of hormones and grade every body accordingly, factoring in age but ignoring all else? We would still have to decide where to draw the line separating "males" and "females" if we wanted to keep those categories, and some people would fall on a side unfamiliar to them until that point.

Come to think of it, why bother labeling males, females, and hermaphrodites? Part of the reason is that it is important to know about one's own body. My eighth-grade health teacher would have reminded me, for instance, that it is important for me to know that, having been identified as a "female," I stand a much greater chance of getting pregnant or developing cervical cancer than those people identified as "male." Given the way I want to lead my life, and given what it means to be pregnant or to have cancer, I find it beneficial to keep such facts in mind and to know as much about them as is useful to me.[12] Similarly, some forms of sexual "ambiguity" signal an underlying metabolic problem that can lead to pain, physical debilitation, or even death. So knowing about the various kinds of sexual development pathways helps doctors and hermaphrodites watch out for and treat certain diseases.

But a large part of the reason for sex-sorting is that we make very important social distinctions based on malehood and femalehood. My mother is my mother because she is a female, and, as a "mother," she enjoys certain social privileges and suffers certain stereotypes to which my father is not entitled or subjected. Similarly, a man cannot compete in an Olympic event designated as being for women, even if his talents are more in keeping with those of women athletes in that sport. Only women are supposed to use the women's locker room at my gym, partly for reasons of privacy, partly for safety, partly out of sexual tradition. A woman cannot be drafted. And so on. It hardly needs expressing that one's identity as male or female seems to matter every day in very important ways.

Hermaphrodites get reduced in number (by being sorted and surgically made into "boy/man" or "girl/woman" types) chiefly because we have many social distinctions that depend on there being (only) two sexes. After even a cursory study of the phenomenon of sex-sorting, one soon discovers that a significant motivation for the biomedical treatment of hermaphrodites is the desire to keep people straight. That phrase—keeping people straight—should be taken figuratively, but literally as well: medical doctors, scientists, hermaphrodites' parents, and other lay people have historically been interested in sorting people according to their sexes to avoid or prevent what might be considered homosexuality. This is perhaps a necessary outcome in a culture that categorizes sexual encounters primarily according to the sex of the partners and tends to favor heterosexuality over homosexuality. For that matter, even people with much invested in the identity of being gay share with those invested in being straight the desire to see others sorted out according to sex. If you don't know who is a male and who a female, how will you know whether what you've got is a case of heterosexuality or homosexuality? Many assume that if we don't keep males and females sorted, at least at some basic level, social institutions that we hold dear—including divisions into heterosexuality and homosexuality, into mothers and fathers, into women athletes and men athletes—will no longer be viable. No doubt this is true. Our social and personal identities will undergo serious challenge or change if we can't or don't sort people according to sex.

So, if we want to sort, what should we employ as the necessary and/or sufficient traits of malehood and femalehood? What makes a person a male or a female or a hermaphrodite? This is the problem. Today my own students, college students in history classes, sometimes in exasperation ask these questions of me at the end of a discussion of the history of sex, as if I am hiding the "real" answer from them. "What *really* is the key to being male, female, or other?" But, as I tell them, and as we shall see, the answer necessarily changes with time, with place, with technology, and with the many serious implications—theoretical and practical, scientific and political—of any given answer. The answer is, in a critical sense, historical—specific to time and place. There is no "back of the book" final answer to what *must* count for humans as "truly" male, female, or her-maphroditic, even though the decisions we make about such boundaries have important implications. Certainly we can observe some basic and important patterns in the bodies we call "male" and the bodies we call "female." And the patterns we notice depend in part on the cognitive and material tools available at a given moment. But the development of new tools doesn't get us closer and closer to some final, definite answer of what it is to be "truly" male, female, or hermaphroditic. Instead it only alters the parameters of possible answers. A hundred years ago we could not point to "genes" in the way we can today, but being able to point to genes doesn't mean that we have found the ultimate, necessary, for-all-time answer to what it means to be of a certain sex. The ultimate decision to define males, females, and hermaphrodites in particular ways necessarily depends all at once on contemporary concepts and available technologies and the tolerance or intolerance of a given definition's larger implications. What it means to be a male, a female, or a hermaphrodite—and how you know you are a male, female, or hermaphrodite, and what will happen to you if you are identified as a male, a female, or a hermaphrodite—is specific to time and place.

THE BODY OF this book forms a history of the biomedical treatment of human hermaphrodites in France and Britain in the late nineteenth and early twentieth centuries. It tells the story of the encounters of two groups of people who met in what seemed to many the most unlikely of circumstances and changed each other's ideas, beliefs, and lives. In the one group were the people like Sophie V., so-called hermaphrodites and other people of "doubtful" or "mistaken" sex whose bodies did not look the way others assumed they should. In the second were the men like Dandois and Michaux, men of medicine and science who grappled with those of the first group theoretically and practically. (I use the phrases "men of medicine and science" and "medical and scientific men" because the scientists and medical practitioners in this history were all men, with the one exception noted in Chapter 3. This point is of obvious significance given that this is a history of the negotiation of the nature of sex.) This book documents one critical moment in the history of sexual identity, a moment when sexual identity, in scientific theory and medical practice, came to be questioned and resolved and questioned again. We see in this one detailed case study of hermaphrodites how and why ideas of sex, gender, sexuality, and identity are formed and changed. Today "sex" is considered a strictly anatomical category; "gender" is used as a category of self-and/or social identification (to say something is "gendered" as opposed to "sexed" implies some doubt that it is "naturally" linked to anatomical sex); and "sexuality" is a term used to refer to sexual desires and/or acts. Although I employ these terms as specified here in the subsequent discussion, these distinctions are not self-evident and were not drawn in the period under consideration. We shall see how and why French and British medical and scientific men who treated hermaphroditism in the late nineteenth and early twentieth centuries developed the specific—and intertwined—ideas they did about sex, sexuality, and gender. We shall also see how, through the handling of difficult problems such as human hermaphroditism, scientists and medical doctors came to be the accepted and sought-after authorities in matters of sex and, more generally, the accepted and sought-after authorities in nearly all matters concerning anatomies and identities.

This is, then, a story about more than hermaphrodites and doctors. It is a story about bodies and beings and what these things have meant and could mean to each other. It is as well a story about a struggle over who would decide what anatomy and identity could and would and should mean to each other. Although in some ways the story that follows is a rich and complicated tale, involving many people of many different backgrounds, experiences, beliefs, and desires, in other ways it is fairly simple, a tale about the rise of science and medicine in the West. It is full of a great variety of people, ideas, and experiences, but it traces throughout two important developments: first, the trend by which, increasingly through time, the Michaux (the biomedical experts) of the world would garner the ability and authority to say what one's flesh did and could mean to one's identity; and second, the trend by which, increasingly through time, the Sophies (the hermaphrodites) of the world would be made less and less visible in the world outside hospitals and clinics and medical journals.

This history, because of its depth and complexities, does not lend itself to a straight chronological telling. Therefore the following chapters are for the most part organized thematically, with important chronological developments noted where appropriate. The first chapter, entitled "Doubtful Sex," introduces the reader more fully to the phenomenon of hermaphroditism, including past and current understandings of it. In this chapter we also gain a better sense of how and why, with the potential for an explosion of "doubtful sex," medical and scientific men in late nineteenth-century Britain and France were driven to seek a stringent, convincing, and permanent definition of "true" malehood, femalehood, and hermaphroditism—why it was decided among biomedical experts by about 1890 to instate what I call a gonadal definition of sex, that is, a rather extraordinary, uniform sex classification system according to which every body's "true" sex would be marked by one trait and one trait only, the anatomical nature of a person's gonads: the ovaries or testicles. If you had ovaries—no matter what else—you would be a woman; testicles, a man; only if both, a "true" hermaphrodite. I therefore call this period the "Age of Gonads."

The second chapter, "Doubtful Status," samples the range of nineteenth-century encounters between hermaphrodites and medical men. It provides an overview of the two groups involved and an account of their multiple interests in each other. Hermaphrodites and the professional men who grappled with them in theory and practice had much to gain from cooperation—not the least of which was the securing of badly desired identities: for hermaphrodites, the identity of one particular sex (which sex depended on the hermaphrodite); and for the medical and scientific men, the role of professional and cultural authorities.

Many historical studies have documented the so-called construction of womanhood and femininity in science and medicine in the nineteenth and twentieth centuries. Too few, however, have looked at concurrent constructions of manhood and masculinity. Here we trace both, because biomedical discussions of hermaphroditism offer a unique window into assumptions about the nature of masculinity *and* femininity. In the third chapter, "In Search of the Veritable Vulva," I survey more encounters like that of Michaux and Sophie, and catalog which behavioral and anatomical traits were believed to be sex linked. I delineate as well the many remarkable disagreements that existed among medical men with regard to the essence and signifiers of femalehood and malehood. Claims about the nature of each sex were not only forced to the surface and voiced in cases of hermaphroditism. They were also discussed, argued, and invented in professional discussions of the cases as medical and scientific men set out on a quest for the "veritable vulva." We see here an active negotiation among groups of professional men about which traits and which body parts would count as essentially or significantly masculine or feminine. A variety of opinions and possibilities emerged.[13]

The fourth chapter, "Hermaphrodites in Love," is devoted to biomedical interpretations of hermaphrodites' sexuality, that is, their sexual desires and acts. In this chapter I demonstrate that the conceptual and practical treatment of hermaphrodites by medical and scientific men was based on the assumption that heterosexuality was the only natural, normal, and moral form of sexuality. Medical men were very disturbed by the conceptual and practical messes hermaphrodites' love lives seemed to cause, and they struggled to bring order to hermaphrodites' sexuality lest the sexual disorder spread. We see here the emergence in medical theory of two unlikely identities, namely the "homosexual hermaphrodite" and the "hermaphroditic homosexual," made possible—even necessary—by the meeting of biomedical theories of sexuality, dominant social mores, and clinical cases of hermaphroditism.

While the third and fourth chapters show how medical men tried to cope with hermaphrodites in practice, the fifth chapter, "The Age of Gonads," focuses on how medical and scientific men also tried to rein in "doubtful sex" in the theoretical realm. I trace here the evolution of a gonadal definition of sex, that is, a definition of "true sex" based on the simple presence of ovarian or testicular tissue. In medical practice sex seemed to slip and slide all over the body; in scientific and medical theory sex came to be limited successfully to tiny bits of tissue. Indeed, this classification system developed in direct response to the exigencies of hermaphroditism and especially in response to the pervasive interest among medical and scientific men in keeping social sex borders clear, distinct, and "naturally" justified. We also see in this chapter how diagnostic technologies enabled and forced certain sorts of redefinitions. Experts had to keep changing their definitions to stay ahead of the possibility of living "true" hermaphrodites.

The decision about which sex to assign a given doubtful subject often involved a struggle among medical practitioners and between medical practitioners and the lay people involved. We will see in the following chapters the diversity of hermaphrodites' experiences and identities, diversity made possible in part by the fact that medicine, science, and anatomy itself had not yet achieved the cultural authority they later would. Many authorities existed in the lives of hermaphrodites in the late nineteenth century, from midwives to grandmothers, to surgeons and physicians, to lovers and friends, to the hermaphrodite him/herself. Today, by contrast, especially in the United States, physicians possess the vast share of the say of what a person's sex is and/or will be. In an epilogue entitled "Categorical Imperatives," I discuss the treatment protocols for hermaphroditism employed in U.S. medicine today, and I raise questions about whether, given the history explored in this book, these practices should be reconsidered and revised. The encounters of hermaphrodites and medical doctors have changed in some aspects, but in many ways they retain troublesome baggage from events of a century past.

With regard to the Epilogue it is worth mentioning that when I began work on this book in 1990, it was my intention to stick very closely to one particular chunk of history, do that historical work as best I could, and leave it mostly up to the reader to draw conclusions about what the history I provided might have to do with related issues. As I shared my findings with others, however, I began to realize that many people were interested to hear how this study might illuminate the treatment of intersexuality in the United States today. Mostly because I discovered that my audiences expected it of me, I took the plunge and started to think and talk about what history can teach us about present-day intersexuality.

As the years went by and my work started to circulate more broadly, I began to meet people who were interested in my work because they themselves had been born intersexual (hermaphroditic). Some of them have become my friends over time, but I feared contact with them at first. That fear arose not out of disinterest in or a lack of sympathy for their experiences, or out of a gut-level fear of meeting living, self-identified intersexuals—or so I told myself—but out of a fear of the demands living intersexuals might make of me and my work, some of which I thought I might not be able or willing to meet. I often wanted to retreat to a paraphrase of *Star Trek*'s Dr. McCoy—"Damn it, I'm a historian, not a doctor or an ethicist or a sociologist!"

But as I talked more to the people I met who had discovered that they had been born intersexual, and as I talked to physicians, friends, and colleagues, I came to a delayed revelation. I had to at least ask living intersexuals for feedback on my work, because my work was in a profound way more about them than about anyone else except perhaps doctors and scientists (with whom, incidentally, I talked to regularly in spite of parallel fears to those named above). I also realized that I had to talk and write about the problems of the current-day treatment of intersexuality. Problems exist, and many are best understood after a careful look at the history of medicine's encounters with intersexuality.

I hope the reader will use the history that follows to think about many broader and related issues. I can say that in researching the history of hermaphrodites for the past several years, I have learned not only about the history of biology, medicine, unusual bodies, and the history of the unexpected and changing answers to my eighth-grade health teacher's question. I have also learned about science and culture, about technology and language, about what the body could mean, does mean, and perhaps should mean, to the life of the person. Without realizing I would, in this sometimes mind-boggling exploration of the experiences of bearded women and menstruating men, I have learned about how and why it is that scientists and medical doctors work to mediate the relationships between our bodies and our selves. I have also learned much about why it is we often look to scientists and medical doctors to read or even to alter our bodies.

The way I see the world has changed drastically. Besides being utterly unable to answer my teacher's question—"How can you tell if you are a boy or a girl?"—with anything but historical stories, I now think about nose jobs, "gay genes," amniocentesis, Huntington's Disease, life support, prenatal medicine, "Siamese" twins, prescription drugs, and my father's portable wheelchair in ways I never could before. I realize how complicated life can be if one has a body or a being that stands out from the rest. And do not we all? Surely in some way every one of us is like a hermaphrodite, a being or a body that won't quite fit the boundaries. So I do not think of this intellectual journey only as a detailed history of sex, gender, and sexuality. I think of it as an exploration of what it means to have a body and a being in this day and age.

CHAPTER 1

Doubtful Sex

In reality, is the subject a boy or a girl?

—Dr. Gaffe, reporting on a case of hermaphroditism in the *Journal de médecine et de chirurgie pratiques* (1885)

Reality is crushing me, is pursuing me. What is going to become of me?

—Abel (née Alexina) Barbin, autobiography (c. 1868)

I wrestled with reality for thirty-five years, Doctor, and I'm happy to state I finally won out over it.

—Jimmy Stewart as Elwood P. Dowd in *Harvey*

THE HISTORY OF HERMAPHRODITISM is largely the history of struggles over the "realities" of sex—the nature of "true" sex, the proper roles of the sexes, the question of what sex can, should, or must mean. We have already seen one such struggle in the form of the clash between Sophie V. and Michaux over Sophie's "true" sexual identity. Although that particular reality-wrestling match happened in Belgium, throughout the late nineteenth century, all over Europe there occurred similar sorts of encounters between hermaphrodites and medical and scientific men. Much recent scholarship has been devoted to documenting the nineteenth-century medical and scientific construction of sex and gender, especially of women and femininity.[1] Such studies show how scientific and medical men used their minds, tools, and rhetoric to build up powerful ideas about the "true natures" (and limits) of femininity and masculinity. My review of medical literature on hermaphrodites supports in part the findings of these studies in that it similarly demonstrates rampant sexing of—that is, attributing of the designation of "male" or "female" to—various body parts, behaviors, and desires by medical and scientific men and lay people of the time. The literature on hermaphroditism also reveals, however, that there was not a single, unified medical opinion about which traits should count as essentially or significantly feminine or masculine. In France and Britain, the sexes were constructed in many different, sometimes conflicting ways in hermaphrodite theory and medical practice, as medical men struggled to come up with a system of sex difference that would hold. Ultimately it was not only the hermaphrodite's body that lay ensconced in ambiguity, but medical and scientific concepts of the male and the female as well. We see here not stagnant ideas about sex, but vibrant, growing, struggling theories. Sex itself was still open to doubt.

Witness now the stories of two nineteenth-century hermaphrodites, one the relatively well known Abel/Alexina Barbin of France, the other a "living specimen of a hermaphrodite" who quietly surfaced in England twenty years after Barbin's death.

Tales of Two Hermaphrodites

In Paris in 1868, during the month of February, a local police commissioner and one Monsieur Régnier, a physician in the employ of the state registry office, were called to investigate the suicide of a young man who had worked as an administrative clerk for the Parisian railroads.[2] When the two officials mounted the stairs to the scene of the death, a squalid, sparsely furnished room on the fifth floor of a boarding house on the rue de l'École-de-Médecine, they found the corpse of Abel Barbin lying across the bed. Barbin had evidently used the small charcoal stove in his room to end his life by carbon monoxide poisoning. A discharge of dark, frothy blood spilled from his still lips.[3] Surveying the lamentable scene, Régnier decided to remove the few articles of clothing in which Barbin had died, and to examine the subject's genitalia. The doctor suspected the young man's melancholy had grown either directly or indirectly from a familiar cause: syphilis.[4] During a genital examination, however, much to his surprise, Régnier discovered not signs of syphilis, but a strange mélange of sexual anatomy: a short, imperforate penis, curved slightly backward and pointed toward what Régnier could only call a vulva—labia minora and majora, and a vagina large enough to admit an index finger.[5]

News of the discovery of a recently dead "hermaphrodite" soon reached the Faculty of Medicine, including the *lauréat* Dr. E. Goujon, who feared that if some member of the faculty did not act quickly, "this observation," this unusual body, "would be lost for science."[6] Goujon quickly mobilized, determined to see the anomaly for himself. He located Régnier and arranged to have the corpse turned over to the faculty for study. While performing the autopsy Goujon took careful notes, for he intended to publish his findings of this strange case.

Before long, however, Goujon discovered his publication would not be the literary debut of this hermaphrodite. Eight years earlier, a doctor in La Rochelle by the name of Chestnet had reported the history and anatomy of this very same "man" in the pages of the *Annales d'hygiène publique* (Annals of Public Hygiene) on the occasion of the revision of the subject's civil status from female to male. Moreover, to the great excitement of all the medical men who would involve themselves in or with the story of Barbin, the subject himself had left behind extensive memoirs, dating from about 1864, or approximately three years after the legal sex revision.[7] In thick, dramatic prose, Barbin had recorded every detail of the trials, tribulations, loves, and losses of her/his tortured soul, as well as her/his persistent conviction that s/he could not go on, that death was imminent. In this autobiography—a text that would be published in 1874 by one fascinated medical doctor as part of a treatise on the "Medico-Legal Question of Identity"[8]—Barbin imagined the day of her/his death:

> When that day comes a few doctors will make a little stir around my corpse; they will shatter all the extinct mechanisms of its impulses, will draw new information from it, will analyze all the mysterious sufferings that were heaped up on a single human being.[9]

It turned out to be an eerily prescient dream.

Twenty-nine years before his piteous death, in the town of Saint-Jean-d'Angély, Barbin was born and christened as a girl. It is not clear whether any questions were raised about the child's sex at that time. We do know that within a few days of the birth, as required by French law, the newborn was presented to the local mayor and civil status registrar so that they might record certain facts: the child's birthdate (November 8, 1838), the names of the parents (Jean Barbin and Adélaïde Destouches), the sex of the child (female), and her name (Adélaïde Herculine Barbin). From her childhood until her legal sex revision, Barbin's familiars knew her as Alexina.

Uncommon anatomy was not the only trouble with which Alexina Barbin was cursed. Her father died when she was a small girl, thus compelling her desperate mother to give up her only child to the care of an order of nuns, the Ursulines of Chavagnes. In the convent boarding school, Alexina enjoyed a stable life and a good education. She was a solid and obedient student, and eventually agreed to become a teacher of girls herself, though she confessed she "was no more flattered by the prospect of being a *working woman*. I believed I deserved better than that."[10] Trained at the normal school of Oléron in Le Château from 1856 to 1858, Alexina gained a position at a small girls' boarding school directed by an established teacher and her sister, the latter a young, unmarried woman by the name of Sara. Sara lived with her mother, Madame P., who provided the new teacher with lodging in their home.

Alexina—an individual of somewhat stocky build, who had, to her dismay, never menstruated—and Sara, whom Alexina saw as a model of feminine grace, grew intimate quickly. Indeed they became nearly inseparable, even going so far as to share a bed regularly. Witnessing their strong attachment, Madame P. scolded them, "You are very fond of each other, and for my part I am very happy that you are; but there are proprieties that must be observed, even among *girls.*"[11] Still, Madame did not suspect the level of their intimacy. Not long after Alexina's arrival, she and Sara had begun having sexual relations. As Alexina remarked, Sara's mother "saw me only as her daughter's *girlfriend,* while in fact I was her lover!"[12]

A combination of a weighty conscience and a painful abdomen finally led the tormented Alexina to a series of priest-confessors and medical men, the result of which was a consensus that Alexina was a man, a male who had been mistaken at birth for a female, and that therefore her legal and public identity ought to be "rectified" to match her "true sex." In 1860 her case was brought before the tribunal of her home district, with the critical testimony entered by Dr. Chestnet of La Rochelle. On June 21, 1860, the register of Barbin's birth was amended. Her sex was changed to male, and her name to Abel. The record now showed she had always truly been a male.[13]

"According to my civil status, I was henceforth to belong to that half of the human race which is called the stronger sex," Alexina—now Abel—noted in the memoirs.[14] Barbin hoped that this new public identity might enable him to marry Sara, a dream the two apparently shared. Such happiness was not to be; the scandal was too great. Instead a family benefactor aided Barbin by securing him a position of clerk to a Parisian railroad. So to Paris he went, where no one would know him and his past. Nevertheless, Barbin could not escape the painful incongruities of his life. Alexina had detested the thought of being a working woman, and now she found herself a working man. He bemoaned his fate: "Reality is crushing me, is pursuing me. What is going to become of me?"[15]

Within a short time, before reaching the age of thirty years, the body of Abel Barbin lay on the dissecting table of Goujon. We do not know precisely what finally drove Barbin to suicide: the lost love? the sudden change in identity? something else instead? In any case, at the autopsy, the doctor saw to it that careful sketches of Barbin's unusual anatomy were made and recorded in the medical literature for any curious medical man to peruse (see Figures 1 and 2). Over the next several decades, many did.[16] There is, however, no known picture of Alexina Barbin's or Abel Barbin's face.[17]

BY CONTRAST, the face of a hermaphrodite who surfaced in England in 1888 was recorded, though perhaps unintentionally, looming as it did in the background of a photograph focused on this subject's uncommon genital conformation (see Figure 3). On Wednesday, April 25, 1888, at a

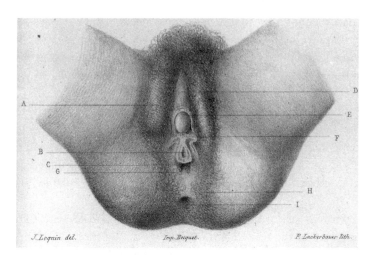

Figure 1 The sexual anatomy of Alexina/Abel Barbin, caudal view. According to Goujon's text, the letters indicate: A. right testicle (left did not descend); B. opening of the urethra; C. vagina; D. "phallus or clitoris"; E. skin which covered the glans; F. orifices of the sebaceous glands; G. orifice of the excretory canals of the sperm; H. hairs; I. anus.

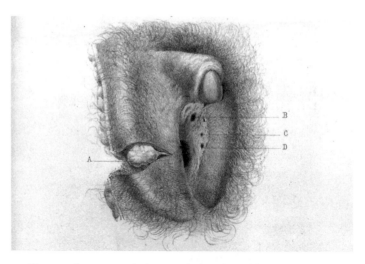

Figure 2 The sexual anatomy of Alexina/Abel Barbin, oblique view. Letters indicate: A. "incision through the skin allowing viewing of the vulvo-vaginal gland and its excretory canal"; B. urethra; C. external orifice of the excretory canal of the vulvo-vaginal gland; D. external orifice of the deferent canal.

regularly scheduled meeting of the British Gynaecological Society, the esteemed gynecologist Dr. Fancourt Barnes, physician to the Chelsea Hospital for Women, the British Lying-in Hospital, and the Royal Maternity Charity, announced that he had "in the next room a living specimen of a hermaphrodite." Barnes added that he used the term hermaphrodite "on general principles, because it is the case of an individual who has been brought up to the age of nineteen as belonging to the female sex, when it is perfectly clear that he was a male."[18] Or was it perfectly clear?

Although a few colleagues—among them the well-known gynecologist Lawson Tait, surgeon to the Birmingham and Midland Hospital for Women, a man like Barnes well versed in hermaphroditism—supported Barnes's diagnosis, several society members in attendance were not so sure. Pressed by the latter to defend his claim, Barnes explained that

> his own reasons for believing the person to be a male were, (1) the appearance of the head, (2) the *timbre* of the voice, (3) the non-development of the breasts, (4) the undoubted existence of a well-formed prepuce and glans penis, (5) the imperfectly formed urethra running down from the tip of the glands [*sic*] and passing into the bladder, (6) the utter absence of anything like a uterus or ovaries, and (7) the appearance of the perineum. The thighs were covered with

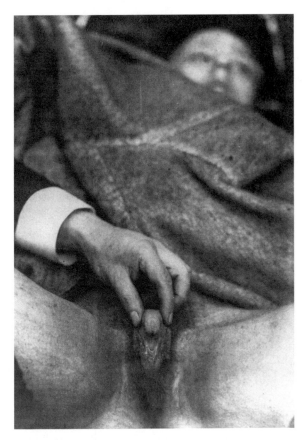

Figure 3 Fancourt Barnes's "living specimen of a hermaphrodite" with the hand of an unknown medical man (possibly Barnes) holding up the phallus.

masculine hairs . . . Lastly, the patient never had the menstrual molimina.[19]

These seemed to Barnes significant and sufficient symptoms of malehood. But if Barnes had come to the meeting thinking that his diagnosis (that the patient was "undoubtedly a man") would be uncontroversial, he was sorely mistaken.[20]

Criticism and controversy began with an objection from Dr. Charles Henry Felix Routh, consulting physician to the Samaritan Free Hospital and founding member of the British Gynaecological Society. Routh insisted it "was by no means clear that it was a man. Such a conclusion was mere guess work."[21] Routh asked whether the rectal examination conducted by Barnes had been sufficient, specifically whether it was enough to have performed a digital exam, or if, rather, Barnes should have tried to "pass his entire hand into her rectum" to search for evidence of a uterus. Furthermore, Routh objected, "Even supposing there were no uterus, the mere fact was no argument against its being a woman." This might simply be an instance of a woman deprived of that organ. Routh was unconvinced. Barnes offered the additional evidence for manhood that two or three years earlier there had appeared a moustache and beard, but Routh protested that "this had absolutely no weight. Many Jewesses had quite a large quantity of both beard and moustache." Beards and moustaches, like absent uteri, were not unheard of in women. In Routh's eyes these were not sure signs of sex.[22]

As consulting physician to the Chelsea Hospital for Women, Dr. James Aveling had to agree with Dr. Routh on the matter of facial hair, for he himself "had often seen it in women just as well marked." Aveling also noted of Barnes's subject that "[t]he face was feminine, [and] the throat was decidedly that of a woman, the *pomum Adami* not being at all prominent."[23] He recommended putting the patient under an anesthetic and performing a bimanual examination "per rectum and through the abdomen."[24] He wanted, if at all possible, evidence of ovaries or testes. Without that, perhaps Barnes had been too quick to conclude.

For his part, Dr. G. Granville Bantock, surgeon to the Samaritan Free Hospital, concurred that "[t]hey all knew how unsafe facial characteristics were as a guide, the dress making such a very great difference." Nonetheless he found the case "clear as daylight." To him the critical signs lay well below the face: "The appearance of the sexual organ was that of a penis, and the remainder of the urethra was to his mind, incontrovertible evidence in favor of its being a male."[25] Still this reasoning did not persuade everyone present. After all, deceptive genitalia were the root of the problem in such cases of "doubtful sex"; it hardly seemed that in cases of ambiguous sex the offending organs should stand as trustworthy witnesses. So the fellows now argued about how much weight ought to be given to *these* signs, or to pubic hair and menstruation, or lack thereof, to the shape of the skull, to balding patterns. Dr. Routh pondered aloud the question of whether a woman without a uterus and ovaries would still be a woman. Piling doubt upon doubt, Dr. Bantock paused to criticize Dr. Aveling's handling of a similar case some years prior, with the result that a subdebate erupted between Bantock and Aveling over what should have been the proper diagnosis in that case.

Finally, in an effort to bring to a close this fractious exchange, the London physician Dr. Heywood Smith "suggested that the Society should divide on the question of sex."[26] Divide these men did, dramatically unable now to decide what they had seen and felt, incapable of agreeing on the nature of sex and its proper diagnosis. The "living specimen" was twice photographed for the record, and apparently then departed with his/her mother, the doubt here left uncomfortably unresolved.

The Intolerance of Hermaphrodites and the Age of Gonads

Many French and English readers of our time are familiar with the tragic story of Alexina/Abel Barbin, thanks mostly to the work of Michel Foucault who, about two decades ago, directed the republication in English and in the original French of Barbin's memoirs. It is an unfortunate fact, however, that Barbin's memoirs are the only known record of their type from nineteenth-century Europe. How much better off we would be in our understanding of sexual identity and hermaphroditism if we could, as Roy Porter suggested in 1985, do a good part of medical history "from below," in this case from a study of the hermaphrodites' viewpoint.[27] How much more complex and rich the history would be if, for instance, Sophie V. and Barnes's "living specimen" had left behind their impressions of their bodies and experiences, especially given that their "true" sexes remained overtly in doubt, while Barbin's was publicly settled.

The absence of documents like Barbin's memoirs cannot justify the far too sweeping conclusions—about scientific and social concepts of sex, gender, and sexuality—that have been drawn by some recent scholars simply from the singular case of Barbin. As we can see from the cases of Sophie V. and Barnes's "living specimen," the lack of documents like Barbin's memoirs does not mean that all knowledge of hermaphrodites and their experiences aside from Barbin's is lost to us. Important information about hermaphrodites' lives and views can be garnered from the sizable body of medical and scientific literature on the subject, keeping in mind that this information is filtered and shaped by contemporary biomedical observers and authors.

Indeed, we can see even from just the three cases discussed so far how variable the encounter of hermaphrodites and medical men could be in the nineteenth century. Sometimes the issue of the patient's "true sex" was resolved, sometimes not. Sometimes the medical men's decisions were heeded, sometimes not. Sometimes there was agreement among the experts on what mattered in finding "true sex," and sometimes not. These are important variations, and if we look at just one case like Barbin's, we lose sight of all the turmoil and negotiation that occurred over doubtful patients and the richness of experience of the hermaphrodites themselves. Some of this variation in case histories, as we shall see, is attributable to cultural differences in France and Britain. This study intentionally looks at hermaphroditism in two nations of the same period in order to explore how—even in fairly similar cultures—the understanding and treatment of hermaphrodites varies according to culture. But these encounters also varied because the individuals in them were of different motivations, convictions, educations, and experiences.

When I set out to do this work, I made a point of trying to locate as many documents as possible about cases of hermaphroditism in France and Britain during the period in which I was interested. I have been able to find roughly three hundred commentaries and accounts of human hermaphroditism published in the scientific and medical literature of France and Britain from 1860 to 1915. For comparison I have also read many others published before and after this period. These collectively refer to approximately two hundred different cases of hermaphroditism, roughly one-half to two-thirds of which seem to have surfaced in France and Britain after 1870 and before 1915. (It is not always possible to tell from the record where or when a case was uncovered or whether it is the same as one reported elsewhere.)

With the exception of Barbin's memoirs, all of the primary source texts I have located were written by medical and scientific men. The lack of documents written by hermaphrodites and the preponderance of men among biomedical authors means that we trace here the development of a peculiarly masculine, biomedical view of masculinity, femininity, and hermpahroditism.[28] By the late nineteenth century, women were starting to make inroads into medical schools, but the fact was that throughout that century in Europe and America women were largely excluded from obtaining medical degrees and licenses.[29] Moreover, as Ornella Moscucci has demonstrated in her study of the nineteenth-century rise of gynecology in England, "opposition to female practitioners was greatest among obstetricians and gynaecologists"—the types of medical professionals most likely to discover a hermaphroditic subject—because obstetricians and gynecologists "had the most to fear from the competition of medical women, for the major argument in favour of lady doctors centered on women's need for medical care which did not violate women's modesty."[30] A woman medical physician or surgeon was therefore uncommon; one who had the opportunity to discover a hermaphrodite even more so.

Of course, during the late nineteenth century many women were still serviced by a different kind of medical woman—a midwife—despite the fact that for over a century midwives had been suffering from the verbal and political attacks and from the professional competition of medical men. Midwives were likely to come upon cases of unusual genitalia with reasonable frequency. However, I have been unable to locate a body of literature on hermaphroditism by midwives of the late nineteenth century. Part of the reason for this may be that midwives did not share the same level of interest in publication that the professionalizing medical men did. Additionally, in searching for relevant documents I was obligated primarily to use indexes compiled by men often disinterested in the writings of midwives. When, in the literature on "doubtful sex" by men physicians and surgeons, the labors of midwives are discussed, they are almost always mentioned in passing only to be condemned as contributing to disastrous cases of "mistaken" sex assignment.

In the last few years of the nineteenth century, there occurred a virtual explosion of human hermaphroditism. Why did this happen? Some people have suggested to me that a significant increase in industrial pollution could have contributed to the apparent rise in numbers of cases of hermaphroditism, but it is very hard to know what, if any, significant material environmental changes might have given impetus to the rise. (The question of the general frequency of hermaphroditism is discussed later in this chapter.) Instead I think it reasonable to credit the steady rise to other sorts of important social changes.

First, because gynecology and access to medical care were on the rise, more and more people in France and Britain were subject to genital examinations. This meant an increase in the number of people considered to be of doubtful or mistaken sex; there was simply more opportunity for doubt to surface because more genitals fell under careful scrutiny. Second, a dramatic increase in the opportunities for reporting of medical findings—there were more and more outlets for medical publications—meant more opportunities for cases to be documented. Relatedly, there spread in consequence a conviction among medical practitioners that hermaphroditism was not all that rare and that practitioners should be on the look out for the phenomenon—and that, when they found it, they should report it.[31] Seek and ye are more likely to find, find and ye shall publish, publish and the door is opened.[32]

But the amount of ink and angst dedicated to hermaphrodites probably also increased among medical and scientific men in the late nineteenth century because that period saw a proliferation of people like feminists (including some women physicians) and homosexuals who vigorously challenged sexual boundaries. These border challenges resulted in a con-comitant reaction on the part of many medical and scientific men to insist on tighter definitions of acceptable forms of malehood and femalehood, and more bodies therefore fell into the "doubtful" range.

Hermaphrodites like Barbin and Barnes's unnamed "living specimen" embodied serious practical problems for the medical and scientific men who grappled with them. This was not a time when most of these men were interested in seeing sexual boundaries blur, not a time when they found terribly amusing the sorts of obvious, profound disagreements about sex the British Gynecological Society had had to suffer in 1888. The late nineteenth century in France and Britain was a time (not unlike our own fin-de-siècle) in which the nature of the sexes stood as a politically charged issue. Many men of science, following in the footsteps of the great Charles Darwin, wrote with confidence and enthusiasm about the natural and profound differences of the male and the female types. The higher the organism on the ladder of life, they said, the more exquisitely differ-entiated the male and female of the species. But such lively, lofty, often poetic discussions of sex were carried on even while—and probably in part because—certain increasingly visible people challenged the borders that were said by most to divide the sexes naturally.

Homosexuals and feminists remained the most publicly problematic sorts. While a number of women sought university educations, medical degrees, and the vote—all supposedly the realm of a few, select men—some men openly loved other men, and some women other women. These "behavioral hermaphrodites" were sometimes considered enough of a nuisance to warrant direct responses by their detractors in the medical and scientific literature. To make matters for scientific sex-dif-ference theorists even more awkward, at the same time, adventurers reporting back from other lands told stories of cultures in which the behavior of men and women seemed not to follow what most Europeans took to be the natural sexual order.[33] Sex seemed to be changeable with conditions of climate or race—now challenged from within and without.

Still, in contrast (and no doubt partly in response) to these challenges, many medical and scientific men in France and Britain vigorously tried to argue and to evidence that the existing social sex boundaries in their cultures reflected and were therefore necessitated by "natural" sex boundaries—that (most) women did what (most) women did because they were female, that (most) men were manly because of their malehood, that to do or be otherwise would not only be unusual, and perhaps immoral, but also unnatural. Never could the two sexes meet in ideas, talents, or roles, for they had parted so long in the past. Men like Patrick Geddes and J. Arthur Thomson, sold on the idea of human evolution, sometimes called on the entire history of life as witness to the necessity of the social sex order. In their popular 1889 tract, *The Evolution of Sex,* Geddes and Thomson declared:

> We have seen that a deep difference in constitution expresses itself in the distinctions between male and female, whether these be physical or mental. The differences may be exaggerated or lessened, but to obliterate them it would be necessary to have all the evolution [of life] over again on a new basis. What was decided among the prehistoric Protozoa cannot be annulled by Act of Parliament . . . We must insist upon the biological considerations underlying the relation of the sexes.[34]

The social sex order had to stand as it did because it was the natural order of things.

Yet in spite of such claims about sex order and profound sex distinctions, as the nineteenth century aged, more and more physicians and surgeons reported in the medical literature cases of anatomically unusual people they called hermaphrodites. Indeed, there seemed to be shocking numbers of anatomical hermaphrodites turning up in the general populace. One French medical man lamented that hermaphrodites seemed to "literally run about on the streets."[35] Hermaphroditism was a sticky problem, one whose possible solution held important ramifications well beyond the life of the individual doubtful patient. So, while deeply fascinated by cases of human hermaphroditism, many medical and scientific men simultaneously expressed disgust at the very idea and resentment at the confusion hermaphrodites caused. In a "Note on Hermaphrodites" in the 1896 volume of his *Archives of Surgery,* Jonathan Hutchinson lamented, "So much of what is repulsive attaches to our ideas of the conditions of an hermaphrodite that we experience a reluctance even to use the word."[36] What was one to do with a person who seemed to be neither or both male and female? What was one to do with the Woman Question, which concerned the proper roles and rights of women, if one could not exactly say what a woman was? How was one to distinguish "normal" (heterosexual) from "perverse" or "inverted" (homosexual) relations if one could not clearly divide all parties into males and females?

Given what Cynthia Eagle Russett has termed "an intense somatic bias" among nineteenth-century scientists—that is, given scientists' deep faith in materialism as the key to truth—the unusual bodies of hermaphrodites presented extremely powerful challenges to biomedical claims about the natural, inviolable distinctions between men and women.[37] Hermaphrodites did not consciously seek to crash sexual borders, but any body which does not clearly fit into the stereotypical categories of male or female necessarily raises questions about the integrity, nature, and limits of those categories. In *Gender Trouble,* a 1990 critical analysis of the relationships of sex, gender, desire, and identity, Judith Butler notes:

> The presuppositions that we make about sexed bodies, about them being one or the other, about the meanings that are said to inhere in them or to follow from being sexed in such a way are suddenly and significantly upset by those examples that fail to comply with the categories that naturalize and stabilize that field of bodies for us within the terms of cultural conventions.[38]

As Butler suggests, anatomical hermaphrodites—however unintentionally—necessarily challenged what it meant to be female or male, woman or man. In doing so they forced observers to admit presuppositions and to make decisions about the categories of female and male—and they forced medical and scientific men to tighten up the borders.

This study takes as its basic starting point the year 1868, that is, the year of Barbin's suicide, and that date is chosen for two reasons. First, the publicity surrounding Barbin's memoirs, life, and death instilled in medical practitioners an appreciation of just how troublesome and urgent—and potentially common—the problem of hermaphroditism was. Medical men understood as never before the dangers of hermaphroditism and the importance of early and accurate diagnoses of an individual's "true" sex. In many ways, Barbin shaped the biomedical treatment of human hermaphroditism for years to come. For example, as in many other instances, the medical man reporting the case of Sophie V. specifically referred to Barbin's suicide as reason to worry about forcing an unfamiliar new sex on a hermaphroditic patient. Meanwhile, Barbin's case, with its scandalous sexual exploits, also pushed many medical men to try to seek out and prevent or end the "unnatural" sexual associations (like Sophie V.'s marriage) caused by hermaphroditism.

Second, and more significant, the year of Barbin's death makes a logical starting point for this history because it was roughly at that time that consensus began to solidify that the single reliable marker of "true sex" in doubtful cases was the gonad, that is, the ovary or testicle.[39] This consensus managed to hold as the rule for assignment of sex until about 1915, that is, the time through which I follow this history. During this period, 1870–1915, what I call the Age of Gonads, scientific and medical men, faced with and frustrated by case after case of "doubtful sex," came to an agreement that every body's "true" sex was marked by one thing and one thing only: the anatomical nature of the gonadal tissue as either ovarian or testicular. Not coincidentally, such a definition virtually eliminated "true" hermaphroditism in theory and practice, even if—probably because—people with challenging bodies kept popping up. Without the material and consequent social problems presented by the hermaphroditic body, this particular construction of "true sex"—namely, that sex is ultimately determined by the gonad—might never have occurred.

Imagine saying that you are female only because you have ovaries, or male only because you have testicles—no matter what the rest of your body or experiences were like. Why pick such a narrow definition of "true" sex? As we shall see, there were many reasons nineteenth-century medical and scientific men were so focused on the gonads: they were partly influenced by trends in contemporary scientific pathology, especially the sort of pathology that used histology as the key to understanding disease; they were partly influenced by trends in contemporary embryology which recognized the differentiation into ovarian or testicular tissue as an embryonically early (and therefore, by the rules of the day, significant) phenomenon; they were partly influenced by trends in evolutionary theories which posited the key difference between males and females to be their reproductive capabilities and roles.[40] Nonetheless, the gonadal definition of sex was not a simple manifestation of trends in related sciences. Without a doubt, scientists and doctors who rallied around the gonadal definition of "true" sex and "true" hermaphroditism did so in part because it meant that nearly every body could officially be limited to one and only one sex. This was critical at a time when the social body itself seemed to be getting too blurry in terms of sexual distinctions. British and French medical practitioners agreed that, as two French physicians writing on the enigmatic problem noted in 1911, "the possession of a [single] sex is a necessity of our social order, for hermaphrodites as well as for normal subjects."[41] The gonadal definition of true sex seemed to preserve, in theory and in practice, a strict separation between males and females, a strict allotment of only one sex to each body—a way to enforce the one-body-one-sex rule.

The Ordering of Hermaphrodites

In this book I use the general term "hermaphrodite" for all so-identified subjects of anatomically double, doubtful, and/or mistaken sex (that is, supposedly mislabeled sex). But I do this not because I think the category of "hermaphrodite" is self-evident or because I think it forms a clearly bounded, ontological category that cannot be disputed. On the contrary, the histories of Barnes's "living specimen," Barbin, and others demonstrate how problematic, diverse, and changeable the categories of hermaphrodite, female, and male are. I use "hermaphrodite" to refer to my historical subjects first because it simplifies my narrative. A single so-called hermaphrodite could, in the scientific and medical literature of the late nineteenth and early twentieth centuries, go from being officially labeled male, to female, to hermaphrodite, to a subject of mistaken sex, to a subject of doubtful sex. So, while I try to note the changes and contradictions in the identities of any given subject, when speaking generally it is simplest to use the word "hermaphrodite" for anyone whose "true" sex fell into question among medical and scientific men. I also use "hermaphrodite" because it was in fact the blanket term commonly used before and during the period of my study for persons suspected of being subjects of double, doubtful, or mistaken sex. The label "hermaphrodite" was sometimes also given to people we would now call homosexuals, transvestites, feminists, and so on, but it was, by the nineteenth century, most commonly reserved for the anatomically "ambiguous" bodies on which I focus. Editors and authors of medical and scientific literature of the nineteenth and early twentieth centuries nearly universally reserved the heading "Hermaphrodite" for reports of cases of doubtful or mistaken sex.

In fact, in spite of repeated attempts on the part of many late nine-teenth-century scientific and medical men to do away with the general, popular, and sometimes vague term "hermaphrodite" in favor of a more rigorous terminology, only since the middle of the twentieth century has a different general term, "intersexed," been regularly substituted for "hermaphroditic" in medical literature. Although these two terms, "in-tersexual" and "hermaphroditic," are used to refer to the same sorts of anatomical conditions, they do signal different ways of thinking about the sexually "ambiguous" body. "Intersexed" literally means that an individ-ual is *between* the sexes—that s/he slips between and blends maleness and femaleness. By contrast the term "hermaphroditic" implies that a person has *both* male and female attributes, that s/he is not a third sex or a blended sex, but instead that s/he is a sort of double sex, that is, in possession of a body which juxtaposes essentially "male" and essentially "female" parts.

Richard Goldschmidt was apparently the first biomedical researcher to use the term "intersexuality" to refer to a wide range of sexual ambiguities including what had previously been known as hermaphroditism. Goldschmidt discussed a variety of "intersexuals" including hermaphro-dites in a 1917 article on "Intersexuality and the Endocrine Aspect of Sex,"[42] and the term "intersexual" slowly gained popularity among medi-cal professionals. Before Goldschmidt, some authors had used the term "intersexuality" to refer to what we would call homosexuality and bi-sexuality, and even Goldschmidt himself suggested that human homo-sexuality might be thought of as one form of intersexuality.[43] (As we will see, the hermaphrodite and the homosexual share a surprising amount of medical history.) Today, however, "intersexual" is used specifically among biomedical professionals to refer to anatomical sex variations considered ambiguous or misleading. Present-day historians sometimes use one term and sometimes the other, as do many of today's activist intersexuals.

Hermaphrodites were not new to the European scene in the nineteenth century. Indeed it was Ovid's myth of the Ur-Hermaphrodite that fixed in the Western imagination the long-standing image of the hermaphrodite as a tragicomic, double-sexed creature. According to the story Ovid told in Book IV of his *Metamorphoses,* the gods Hermes and Aphrodite, them-selves "the embodiments of ideal manhood and womanhood," together had a son named Hermaphroditos after his parents.[44] Hermaphroditos was, like his father, a beautifully formed male, and he might have re-mained so had not the nymph Salmacis one day chanced upon Hermaphroditos while he bathed in her fountain. Having seen his potent form, his remarkable beauty, so badly did Salmacis wish to be united with Hermaphroditos that she begged the gods to join them together: "Oh gods may you so order it, and let / no day take him away from me or me from him."[45] The gods, with their strange sense of humor, took her literally. Hermaphroditos's and Salmacis's bodies were united permanently to become one—one Hermaphrodite—a being made both fully man and woman, "mingled and joined . . . / just as, if someone puts branches through a tree's bark, / he sees them joined as they grow and [mature] together."[46]

Both visions of the hermaphrodite—as an in-between sex and as a double sex—can actually be found as far back as the days of ancient Greece. Joan Cadden has traced the conception of hermaphrodites as in-between sexes back to the Hippocratic writers. The Hippocratic paradigm assumed that sex existed along a sort of continuum from the extreme male to the extreme female and that the hermaphrodite therefore was s/he who lay in the middle. By contrast, later thinkers formed from the writings of Aristotle a different, equally persistent tradition that imagined hermaphrodites to be doubly sexed beings. That tradition specifically held that hermaphrodites had extra sex (genital) parts added on to their single "true" sexes via an excessive generative contribution of matter on the part of the mother. Hermaphroditism in this Aristotelian line of thinking was therefore like extra toes or nipples, in that it represented an overabundance of generative material.[47] Cadden shows that followers of both traditions can be found for centuries beyond ancient Greek times.

Indeed, stories of hermaphrodites are sprinkled throughout virtually every era of recorded history. Sometimes, as in Ovid's case, the hermaphrodite was employed in literature as an allegorical character. In the twelfth century, for instance, John of Salisbury used the image of the hermaphrodite to talk about the strange, double-natured position of a court philosopher, a position which was, according to John, plagued by inherently contradictory (hence "hermaphroditic") loyalties.[48] But we also find references to nonallegorical—that is, living—hermaphrodites in practically every historical era. Some hermaphrodites in Roman and medieval times may have been put to death, as apparently were all kinds of "monstrous" beings. The reasoning behind these murders held that the "monster" was surely a supernatural portent, a messenger of evil, a demonstration (note the shared root with "monster") of bad happenings, and that as such it deserved and even required prompt annihilation.[49] Of course, we must remember that later claims regarding the barbaric treatment of "monsters" by earlier peoples often came during later people's claims to great progress and moral superiority, so they may well have been overestimated. Some hermaphrodites may have been tolerated, but probably only as long as they chose and stuck to a single male or female sexual identity.[50] Hermaphroditism, it seems, has always been a relatively risky identity.

Lorraine Daston and Katharine Park noted in a 1995 study that, in early modern France, scientific thinkers developed a pervasive and singular fascination with hermaphrodites.[51] Men of the sixteenth and seventeenth centuries recorded a dazzling array of opinions and interpretations of hermaphroditism, and Daston and Park suggest that this relatively abundant interest in hermaphrodites arose out of a "sexual anxiety" fed by increasing concerns over transvestitism, sodomy (especially between women), and the possible transgression of other social sex roles. (Interest in hermaphroditism seems almost always to wax with public challenges to sex roles.) In early modern France, hermaphrodites counted chiefly as preternatural beings, that is, "rare and unusual, outside the ordinary course of nature, but in principle fully explicable by natural causes." No longer understood primarily as supernatural portents, hermaphrodites and other "fantastic" creatures became "marvels, not miracles."[52] The fate of hermaphrodites in the cultural scene depended on the nature of others' understanding of the hermaphrodites' origins and actions. While the "preternatural" anatomical hermaphrodites could be explained, understood, and given an ordered place in nature and perhaps as well in the social setting, "artificial" hermaphrodites (those created through human deceit or invention) and "unnatural" hermaphrodites (those whose sexual behavior seemed to go against the "nature" of their "true" sexes) were often persecuted.

By the early nineteenth century, an entire field dedicated to the scientific study of "monstrosities" had emerged. Isidore Geoffroy Saint-Hilaire (1805–1861), a French anatomist, coined the name teratology for the new field. Geoffroy and his cohorts laid out an ambitious goal for the discipline, namely, the exploration of all known and theoretical anatomical "anomalies" and the explication of those anomalies within a single "anatomical philosophy" which would at once describe, explain, and predict all normal and abnormal forms. "Nature is one whole," Geoffroy confidently declared, and all "monsters," including the hermaphrodite, were therefore part of nature. Thus the hermaphrodite—like many other previously supernatural or preternatural phenomena—came to be fully "naturalized," so that by the early nineteenth century, hermaphroditism was understood by scientists and medical men as a phenomenon to be fully explained by the natural sciences, one existing within the realm of natural law.

More specifically, the development of the hermaphrodite and other "anomalous" beings was now to be explained in terms of variations of normal development.[53] Doctors and others settled on the conviction that, in the case of hermaphroditism, the ultimate "explanation is to be found in the embryology of the genitals"; the hermaphrodite was just a would-be male or female gone wrong in the womb.[54] It seemed one had only to understand normal embryological sex development to be able to come up with explanations for abnormal sex development. Nineteenth-century anatomists knew that future-female and future-male children began as embryos with the same basic parts, but that those parts developed differently in utero and after. By the 1870s all authorities in embryology agreed that, in the female fetus, the two proto-gonads took one path to become ovaries, and in the male fetus, they followed another and became testes.[55] In addition, they knew that ultimately-male and ultimately-female fetuses both began with Mullerian and Wolffian systems of proto-organs internally. In the female, however, the Wolffian system atrophied and the Mullerian system evolved to form "female" internal organs, including the fallopian tubes, uterus, and vagina. In the male, the Mullerian system atrophied and the Wolffian system evolved to form "male" internal organs, including the deferent canals and the prostate.[56] Knowledge of such common developmental pathways made it possible to explain, for instance, how a "true male" could seem to have developed an otherwise inexplicable vagina or uterus.

With regard to the external genitalia, medical and scientific men of the nineteenth century saw the human male genitalia as a more elaborately developed version of the human female genitalia. The female seemed to represent simply a lower degree of external development in this sense. This assumption—that females were actually underdeveloped males—was based in part on a long-running anatomical and philosophical tradition that stretched at least back to the work of Aristotle. Aristotle had described women as monstrous, underdeveloped men, "undercooked" for lack of an essential heat. The male heat, by contrast, cooked the male, pushing out the phallus—understood to be the vagina in the woman—and the gonads.[57] But in the nineteenth century, this theory of women as less-developed men was also based on the fact that the external conformation of young embryos seemed to observers to be nearer the ultimate feminine than the ultimate masculine state, and on the observation that masculine genitalia apparently continued to develop after the point at which the development of feminine genitalia stopped.[58] So, for instance, embryologists and anatomists noted that the clitoris in the female is relatively small, but its homologue (developmental counterpart) in the male, the penis, grows comparatively large.[59] Similarly, it seemed to nineteenth-century observers that proto-labia formed in both the male and the female fetus, but in the male, development continued such that the labia joined to form a scrotum.[60] It appeared, therefore, that if a female developed too much, she would look masculine or hermaphroditic, and that if a male developed too little, he would look feminine or hermaphroditic.[61]

Ultimately the work of Geoffroy Saint-Hilaire and other modernist teratologists resulted in what several historians of teratology have labeled the "domestication of the monster." This was the process by which the extraordinary body ceased to be truly extraordinary—that is, outside the realm of the natural—the process by which the teratological was stripped of its wonder and made simply pathological. "Domesticated within the laboratory and the textbook," writes Rosemarie Garland Thomson, "what was once the prodigious monster, the fanciful freak, the strange and subtle curiosity of nature, . . . [became] the abnormal, the intolerable."[62] Elizabeth Grosz joins with Thomson in remarking on the paradoxical fate of the modern hermaphrodite and other contemporary "monster/freaks," namely, to finally be, in "medical discourse and practice" made the subject of "simultaneous normalization and pathologization." Just when (and because) the hermaphrodite was normalized via the alliance of its development with the development of "normal" creatures, simultaneously was it declared unacceptably pathological because of its variation from those very "norms."[63] So, while Geoffroy welcomed human "monstrosities" into the natural realm of the living, at the same time he hoped the science of teratology would have a very useful application: the prevention of future human "monstrosities."[64]

Current-Day Explanations and Typing

Embryological ideas about the parallels in male and female development are still used by biomedical researchers to explain human hermaphroditism, although today experts have added elaborate understandings of the role of genetics and endocrinology to their explanations. Philosophers of hermaphroditism like to recall that Ovid's Hermaphrodite was a sort of "Siamese" twin, a double body at once completely male and completely female with all the attendant parts and traits. But gods and nymphs aside, the human body labeled "hermaphroditic" presents not all the so-called female and male parts, but rather an unusual mix or blend of parts (depending how you think about and see things) as in the cases of Sophie V., Barbin, and Barnes's "living specimen." The organ located where the penis or clitoris is usually found might look like the "wrong" organ (that is, the organ common to the sex "opposite" to which the person is thought to belong), or like something in between the two, or not especially like either. The genitalia may appear to be of the female type, yet the labia might contain testicles. Or the genitalia may look mostly male, but include a seeming vagina. The appearance of the genitals might also change over the course of the lifetime. Change in sexual anatomy—for instance, relative growth of some parts, the growth or loss of hair on other parts—is the rule for most of us, but in some hermaphrodites, genitalia have also been observed to undergo unusual transformations from a more female-like to a more male-like conformation, and vice versa.

Today in the medical literature human hermaphroditism is divided into three basic theoretical types: male pseudohermaphroditism, female pseudohermaphroditism, and true hermaphroditism. Typing in this theoretical taxonomy dates back to the Age of Gonads and so is based primarily on the anatomical structure of the gonadal tissue as was done in the late nineteenth century. Thus if an "ambiguous" individual has testicular tissue only, s/he is technically categorized as a male pseudohermaphrodite; if ovarian tissue only, s/he is categorized as a female pseudohermaphrodite; if s/he has one or more ovotestis, that is, an organ with both ovarian and testicular attributes, s/he is categorized as a true hermaphrodite. The tissue in question need not be functional in any sense. It is only the anatomy of the gonads that is used for this theoretical typing.[65]

Given that medical theory still categorizes hermaphroditisms as pseudomale, pseudofemale, or true on the basis of the anatomy of the gonadal tissue, one might think that we are still in an Age of Gonads. But after about 1915—what I mark as the real end of the Age of Gonads—medical practitioners came to a general agreement that the gonads should not in practice be very important to sex assignment. Today in the

United States, for example, sex assignment is based largely on a concern for penis functionality in males and for reproductive functionality in females. (The Epilogue discusses this at length.) Gonads no longer have the kind of tremendous power they had from the time of Barbin's death to the time of World War I.

Today, "true hermaphroditism"—cases in which an individual has one or more ovotestes—is considered "extremely rare," but it is hard to know how many cases go undiagnosed.[66] Biomedical experts are also not sure how or why a proto-gonad would develop both ovarian and testicular attributes. The vast majority of "true hermaphrodites" seem to have an XX chromosomal basis, although a small percentage exhibit XY chromosomes and a very few have some cells showing XX and others showing XY, a phenomenon known as "chimerism" which probably occurs when two early embryos (one XX and one XY) fuse together to form one individual. The genitalia of true hermaphrodites sometimes look "typically" male, sometimes "typically" female, and sometimes something else.[67] This today is the least well understood type of hermaphroditism.

"Female pseudohermaphrodites" are by contrast much more common, perhaps "accounting for about half of all cases of ambiguous external genitalia."[68] In cases of female pseudohermaphroditism, the individuals in question by definition have ovaries, and they also exhibit an XX chromosomal basis. The external genitalia look "masculinized" because the children are exposed in the womb to relatively high levels of androgens, a type of hormone important in sex-organ development. Sometimes the supposed-clitoris looks and acts more like a penis, sometimes the labia join to look like a scrotum, and so on. Meanwhile, upon examination the internal organs appear "typically" feminine because the developmental anomaly does not profoundly affect internal development.

According to today's expert understanding of female pseudohermaphroditism, the "masculinizing" phenomena of female pseudohermaphroditism can occur for at least three reasons. First, a tumor on the pregnant mother's suprarenal gland can produce excessive amounts of androgens that effect a more "male" type development of the "female" child's genitalia. This seems to be a rare cause. Second, administration of androgenic hormones to a woman pregnant with a "female" fetus can cause "masculinization." Pregnant women have sometimes been given these hormones by their doctors to prevent miscarriage, and there may also be environmental toxins which can cause similar, externally induced "masculinization." Third, "masculinization" in a "female" fetus can be caused by a condition known as congenital adrenal hyperplasia, or CAH. This seems to be the most common cause of female pseudohermaphroditism today. In CAH, the fetus again has an XX chromosomal basis and ovaries, as in all cases of female pseudohermaphroditism, but the production of relatively large amounts of androgens by the adrenal glands of the fetus results in a rather "male"-like development of the external genitalia. Female pseudohermaphroditism can range from very dramatic cases in which the child looks very much like the typical male (some children go years without being diagnosed) to relatively less dramatic cases in which the clitoris is a bit enlarged but most everyone assumes the child must be a girl.[69]

There are at least two major forms of male pseudohermaphroditism. In both of them, by definition, the individuals in question have testes and an XY chromosomal pattern. The first type is known today as testicular feminization syndrome or androgen insensitivity syndrome (AIS). In AIS, the testes produce the usual androgens effective in male development. The body lacks a key androgen receptor, however, and so the body cannot "hear" or "read" the androgen ("masculinizing") messages. Therefore, rather than developing along the typical masculine developmental pathway, the tissues develop along more "feminine" lines. The genitals look quite feminine, with labia, a clitoris, and a vagina (usually relatively short). At puberty the body develops still more along the more "typical" feminine pathway because, although the testes produce more androgens, the body still cannot "hear" them.[70] Breasts fill out, the hips grow rounder. Generally AIS individuals do not develop very much noticeable body hair, and they grow tall with long arms and legs. Indeed, with these features—tall, smooth-skinned bodies with rounded hips and breasts and long limbs—they seem to fit the dominant feminine ideal in the United States today better than most medically "true" females. (There is a persistent rumor that many "female" high-fashion models are testicularly feminized males, but I do not know if this is true, and I have been unable to locate the ultimate source of the rumor.) Individuals with AIS often do not know about their condition until they go to a gynecologist when they fail to menstruate at puberty, and, even at that time, many are not told their true diagnosis, but are simply told they can only become mothers via adoption.[71] AIS patients are frequently told by their doctors that their ovaries have been removed, when in fact they had no ovaries and it is their testes that have been removed. (Their testes are removed because they have a relatively high chance of becoming cancerous, and, since the testes are often undescended, such cancers can easily go undetected until they have reached the point of being terminal.)

The other major type of male pseudohermaphroditism is known as 5-alpha-reductase (5-AR) deficiency. This is one of the most striking forms of hermaphroditism because it results in an apparent female-to-male transformation at puberty. During fetal development the "male" child's testes produce testosterone. But in order for the developmental "message" of the testosterone to be "heard" in the child, the tissues must have the enzyme 5-alpha-reductase, which converts the testosterone into "readable" dihydrotestosterone. If it is lacking, as it is in cases of 5-AR deficiency, the fetus will develop female-like genitalia. Therefore 5-AR deficient individuals are born with feminine-looking genitalia, including generally a short vagina and apparent labia and clitoris. At puberty, however, the testes of these individuals produce more testosterone, and for the pubertal changes to occur the body doesn't need the converting work of the 5-AR enzyme. So now the testosterone messages *are* read, and "masculinizing" puberty occurs. The body grows taller, stronger, more muscular, usually with the addition of significant body and facial hair but with no breast development, and the voice drops. Often at this time the testes descend into the assumed-labia, and the penis/clitoris grows to look and act more like a penis.[72]

These are the types of hermaphroditism generally discussed under the heading of "intersexuality" or "(pseudo)hermaphroditism" in medical textbooks today. However, "ambiguous" genitalia can result from other conditions besides those mentioned above. In Klinefelter's syndrome, for instance, the presence of several X chromosomes and one Y chromosome sometimes results in sexual "ambiguity."[73] We should also note that not all the causes of intersexuality are known or understood and that, moreover, many babies are born with relatively unusually formed genitalia but are not categorized by their physicians as "ambiguous" or "intersexed." Unquestioned females, for instance, are sometimes born with relatively large clitorises, and a large number of male babies—perhaps one in every one or two hundred—are born with hypospadic penises. Hypospadias is a condition in which, in a person with male-type genitalia, the urethra exits some place other than the tip of the glans. The penis may also be relatively small, and it may be bound-down, pointing backwards, to create a condition known as chordee.[74] Today operations are often performed very early on hypospadic penises, in part because it is considered very important that a boy be able to urinate standing up, and relatively large clitorises are often surgically reduced because it is generally considered inappropriate for girls to have large clitorises. While these cases may not officially be labeled cases of intersexuality, the genitalia are treated as cosmetically troublesome.

The Question of Frequency

Many people are understandably curious to know how common the phenomenon of "intersexuality" or "ambiguous genitalia" is. But for several reasons it is almost impossible to provide with any confidence an overall statistic for the frequency of sexually ambiguous births.

First, even the experts today cannot agree on how common various intersex conditions are. For instance, consider statistics given with regard to congenital adrenal hyperplasia. Three recently published, well-respected medical texts give the frequency of CAH alternately as 1 in 60,000 births, 1 in 20,000, and greater than 1 in 12,500.[75] Which statistic to trust? Similar variations are present in estimates with regard to "intersexual" conditions besides CAH. The fact that, even in the most recent medical literature, there exists significant disagreement about the incidence of various intersex conditions makes it difficult to estimate the overall frequency.

Second, and relatedly, part of the problem in determining the frequency of intersexuality or hermaphroditism is a problem inherent in all statistics; there is always a question of whether or not a given sample from which one estimates the frequency of an event represents the larger population in which one is ultimately interested. For instance, when we try to estimate "frequency of intersexual births," are we seeking to estimate it for the entire human population? Since when? Ending when? Only for the United States today? Or for Victorian Britain? Obviously the numbers would differ depending on the parameters, again for several reasons.

Some forms of intersexuality, such as 5-AR deficiency, seem to have a strong genetic component. It seems that 5-AR deficiency can result from a "spontaneous" mutation, but that it can also result from an inheritable genetic sequence. In fact, we know that several populations, through isolation and intermarriage, have come to manifest a much greater frequency of 5-AR deficiency than other populations. One such population exists in a rural area of the Dominican Republic and another among the

Sambia people of Papua New Guinea.[76] Should we include or exclude such clusters of intersexuality in our "overall" estimate of frequency of intersexuality—or even in our estimates of the frequency of 5-AR deficiency alone? Neither including nor excluding cluster cases in our count seems to be satisfactory.

Similarly, we know that certain kinds of "environmental" factors can influence sexual development. For instance, in the 1960s, many pregnant women in the United States were given hormone treatments to prevent miscarriage, and these treatments resulted in a high number of CAH-like conditions. In other words, as in the case of 5-AR deficiency mentioned above, we can find "clustering" of CAH-like cases in the late twentieth-century United States, this time owing to local environmental factors. So even if we could come up with a reliable frequency count for CAH-like conditions in the United States in the 1960s–1990s, that frequency may not be anything like the frequency of CAH-like conditions in other periods and places. The frequency of various conditions is likely to vary quite significantly according to time and space because of local variations in the factors that influence sexual development.

Third, in trying to estimate the overall frequency of sexual ambiguity or sexual anomalies, we encounter the enormous problem of what exactly to count toward such a statistic. Take, for instance, the problem of hypospadias. Like CAH, hypospadias may but does not necessarily result in confusion and disagreement about to which sex an affected person should belong. Similarly, as mentioned above, Klinefelter's syndrome may result in so-called sexual ambiguity, but does not necessarily. Should the overall frequency of hypospadias, CAH, and Klinefelter's syndrome be counted into a general statistic of the frequency of sexual anatomical ambiguity? Or should we try to count only those cases which result in "confusing" or "mistaken" sex? Neither approach seems reasonable. The first would seem to result in too high a count for our purposes, because it would include cases in which the sex was never considered "ambiguous." The second would factor in unintended variables, like the frequency of accurate diagnosis and reporting of hypospadias, CAH, and Klinefelter's.

One might argue that we should simply not lump together all of these various conditions—that we should consider each condition or even each case separately. Yet the phenomenon called alternately "doubtful sex," "ambiguous sex," "hermaphroditism" and so on has been and continues to be treated as a singular sort of problem in medicine and culture, and so should be treated historically as such. Even if we can't find a for-all-time "scientific" way to define intersexuality, that does not mean it does not exist as a very important cultural phenomenon.

In conclusion, it is not possible to provide with any great certainty a statistic of the frequency of births in which the child's sex falls into question. This problem highlights an important point of this book and especially this chapter, namely, that such a statistic is always necessarily culture specific. It varies with gene-pool isolation and environmental influences. It also varies according to what, in a given culture, counts as acceptable variations of malehood or femalehood as opposed to forms considered sexually ambiguous. And it varies according to what opportunities there are in a given culture for doubts to surface and be articulated on record. A culture in which genitalia are covered up, rarely examined, and not discussed would likely be a culture in which there were fewer cases of so-labeled ambiguous sex. A culture in which big clitorises or small penises are considered unacceptable would likely be one in which there were more cases of so-labeled ambiguous sex. Frequency is specific to particular cultural spaces.

It remains true, nevertheless, that the frequency of births in which the child exhibits a condition which today could count as "intersexual" or "sexually ambiguous" is significantly higher than most people outside the medical field (and many inside) assume it is. Curiously, few medical texts offer any estimate of the frequency of "intersexuality" even though many treat its various forms under that single heading. One 1993 gynecology text does offer the claim that "in approximately 1 in 500 births, the sex is doubtful because of the external genitalia," but unfortunately that author does not say how she reached that conclusion.[77] When I am pressed for a rough statistic, I suggest that today, in the United States, probably about one to three in every two thousand people are born with an anatomical conformation not common to the so-called typical male or female such that their unusual anatomies can result in confusion and disagreement about whether they should be considered female or male or something else. Anne Fausto-Sterling, through recent research, estimates the incidence of intersexed births to be in the range of 1 percent, although Fausto-Sterling warns that that figure "should be taken as an order of magnitude estimate rather than a precise count."[78] (In other words, the number might be closer to one in a thousand.) Fausto-Sterling's estimate comes from a careful study of the current medical literature, and after a review of her preliminary findings, I find her compilation of the data convincing, with all the provisos given above taken into account.[79]

It seems quite evident, then, that in the United States today the anatomical condition of intersexuality is about as common as the relatively well known conditions of cystic fibrosis (roughly one in two thousand "Caucasian" births is of a child with cystic fibrosis) and Down syndrome (roughly one in eight hundred live births).[80] Given roughly four million births per year in the United States, that means there are now several thousand medically defined "intersexuals" born in the United States each year, and there are now living in the United States tens of thousands of people who were born to be labeled "intersexual," although many of them probably don't know it. (This is largely because doctors avoid using the terms hermaphroditism or intersexuality around intersexuals and their parents.) Lay people do not hear about intersexuality much in the United States today because such conditions are often considered traumatic, scandalous, and shameful, so few people talk about such births when they happen. Our system of birth registry also covers up the phenomenon: every child, no matter how "ambiguous," gets recorded as male or female very quickly. Nevertheless, intersexuality is becoming more well known in the United States as intersexuals make themselves known and as the phenomenon begins to show up in various pop-culture venues such as the television drama "Chicago Hope," in which a baby was born with AIS, and the feature film "Flirting with Disaster," in which one character (notably the only self-identified bisexual) mentions that he was born with a hypospadic penis.

Body Painting

The latest opinions of scientific researchers are important, of course, to our understanding of intersexuality. But at the same time, human sex has never been and will probably never be a merely academic matter. Ideas about sex, however theoretical, carry with them tremendous ramifications. Consider our subject period: in the immediate sense, a medical practitioner's opinion with regard to the "true sex" of a given patient could shape, in the late nineteenth and early twentieth centuries, at the very least his view of the patient and at most the patient's whole life, and perhaps even death, as in the suicides of Barbin and other hermaphrodites. Imagine what depended on the determination of one's "true" sex: name, dress, education, occupation, possibilities for enfranchisement and conscription, and status as wife or husband, maiden or bachelor, widow or widower, as sexual "normal" or "invert" (what we would call heterosexual or homosexual). Every body that slid through the divisions weakened those boundaries. If strict sex borders were to be maintained in the culture, sex had to be maintained and controlled in the surgical clinics and in the anatomy museums and in every body.

The demands put on the hermaphroditic body therefore are many, as many agendas—scientific, medical, personal, national, professional, moral, and political—meet. This is perhaps inevitable, for in any human culture, a body is never a body unto itself, and bodies that openly challenge significant boundaries are particularly prone to being caught in struggles over those boundaries. Rosemarie Garland Thomson has argued, "By its very presence, the exceptional body seems to compel explanation, inspire representation, and incite regulation. The unexpected body fires rich, if anxious, narratives and practices that probe the contours and boundaries of what we take to be human."[81] Mary Douglas suggested in her study of bodily pollution and taboo that these bodily divisions can in fact be made on virtually any body, extraordinary or not, for "the body is a model which can stand for any bounded system. Its boundaries can represent any boundaries which are threatened or precarious. The body is a complex structure. The functions of its different parts and their relation afford a source of symbols for other complex structures."[82]

So it was that, by the late nineteenth century in France and Britain, the hermaphroditic body became a symbol for many conflicts, as lines were repeatedly painted on it dividing the masculine from the feminine, the normal from the abnormal, the real from the apparent, those with authority from those without. More than any other conflict (and there were many involved), the battle over the "nature" of males and females affected the biomedical treatment of hermaphrodites. Hermaphrodites were repeatedly construed by medical and scientific men so as to reinforce primarily what they threatened most: the idea that there was a single, knowable, male or female "true" sex in every human body.

This idea—that the hermaphrodite's body gets caught up in cultural wars and has painted on it the fronts of those wars—is not novel to our time. In 1899 Xavier Delore published in the *Écho médicale de Lyon* a lengthy study of the cultural-historical, embryological, and evolutionary stages of hermaphroditism, and in his research, the Lyonnaise medical doctor was struck as we are by how, historically, the hermaphrodite had "so strongly impressed public opinion that its history reflect[ed] the prominent traits of the various societies" in which it had been found. But

Delore was convinced that by 1899 the hermaphrodite had finally ceased to be a mythological being, the kind constantly used for the justification of "irrational" and "immoral" beliefs and acts. The hermaphrodite, Delore insisted,

> now no longer has adoring admirers who render it a cult; it is no longer the object of those gracious legends of Greek mythology; it is no longer sung of by the poets of Rome; philosophers no longer utilize its defective structure in order to palliate, by insidious sophisms, the licentiousness of their morals; . . . theologians will no longer use it in order to elucidate ridiculous heresies; but it also will no longer engender that sentiment of horror which unenlightened civilizations drew from it, [only] to exterminate it as a monster!

Now, Delore claimed, the hermaphrodite had become simply, perhaps pathetically, merely "a scientific matter and a degraded organism." As such it had become "the domain of the medical doctors. It [was] incumbent upon them to reconcile its interests with those of society, in the midst of which they [would] mark for it its true place."[83]

Ironically, even while Delore recognized the mapping of many belief systems and cultural agendas onto the body of the hermaphrodite throughout history, he assumed that this practice had essentially ended when the hermaphrodite became an object of modern medical and scientific study. Yet Delore himself readily admitted the exigencies of his own society and profession in matters of hermaphroditism, and the social needs to which Delore alluded necessarily shaped contemporary representations and treatment of hermaphrodites.

Of course the hermaphrodite did not cease to be a site of cultural conflict and cultural demands—of cultural body painting—when it became an object to be handled by science and medicine. It was only the type and status and media of the painters that changed.[84]

CHAPTER 2

Doubtful Status

How am I glutted with conceit of this!
Shall I make spirits fetch me what I please,
Resolve me of all ambiguities,
Perform what desperate enterprise I will?

—Dr. Faustus in Christopher Marlowe, *Dr. Faustus* (1604)

A PERSON TODAY coming upon the 1888 photograph of Barnes's "living specimen" (Figure 3) is likely to be taken aback. But, at least at first, it is not the unusual genital conformation that strikes one. Indeed, the genitals are rather hard to see in this aged picture. Rather, it is the whole scene—the hermaphrodite lying prone, his/her face in a blur, the hand of a man reaching in and holding up the phallus—that might surprise a current-day viewer. Yet this picture is not unusual in comparison with other photographs of late nineteenth-century hermaphrodites. In such pictures the faces of hermaphrodites were often revealed, intentionally or unintentionally (see Figure 4). And photographs of hermaphrodites at this time also tended, like the one in Figure 3, to be somewhat murky. Often sketched diagrams accompanied or were substituted because such sketches made more clear what the parts under consideration looked like. And while today we see relatively few doctors' hands in anatomical illustrations, they are common in photographs and drawings of hermaphrodites from the late nineteenth and early twentieth centuries. (See Figure 5.)

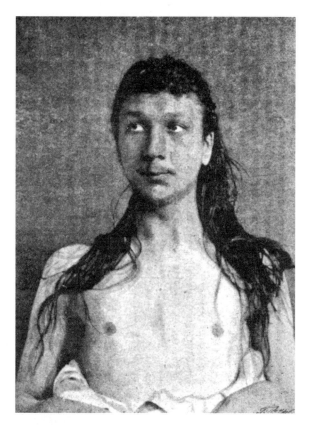

Figure 4 Eugénie Rémy, a discovered case of "mistaken sex." Rémy was thought at birth to be a girl, but several French medical men later decided she was actually a man. Her sexual exploits with her female co-workers bolstered this conviction.

In a way, the reaching-hand motif—startling as it is to late twentieth-century eyes—serves to remind us of two things. First, we see that in all the cases of hermaphrodites under consideration here, at least two persons squeeze into the frame of study, the hermaphrodite and the biomedical observer. As is so well represented in the photograph of Barnes's specimen, however, while the hermaphrodite's ambiguous anatomy and the role of the biomedical observer come relatively easily into focus, the identity and the mind of the hermaphrodite (represented in Barnes's portrait by the blurry face) do not. Second, the hand motif serves to remind us that the meaning and representation of anatomy is always culture specific. In present-day anatomy texts, it seems that the anatomist disappears—that the portrait is self-evident. This is why the reaching-

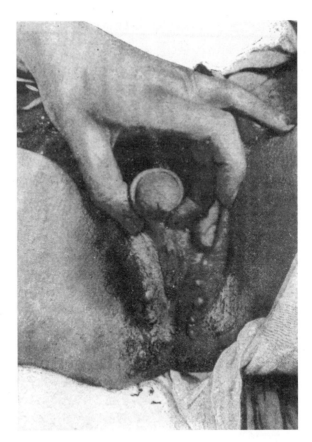

Figure 5 The genitals of Eugénie Rémy, with a hand (presumably of a medical man) holding up the phallus for display. The inclusion of reaching hands was not an artifact of photography; they were often included in sketches as well.

hand motif strikes us now. But in a way present-day texts are much more deceptive than the graphic pictures of a century ago. The century-old illustrations never allow us to forget that there is a "hand" guiding any given image, even if that "hand" is invisible.

Late nineteenth-century medical and scientific men who studied hermaphrodites tended to view their subjects, dead or alive, as "specimens" for study; recall, for instance, that Barnes introduced his case as one of a "living specimen of a hermaphrodite." Ironically, the emerging tradition in the nineteenth century of respecting patients' privacy also served to depersonalize them in the medical realm, often making them literally anonymous. For instance, by the 1890s, hermaphrodites were rarely described in the medical literature using their last, or even their first, names. Now they became "S.B." or "L.H." or, at most, "Sophie V." Meanwhile some investigators began to photograph subjects after placing a black bag over their head (see Figure 6)—the precursor to the practice of placing a black bar across a subject's eyes in later medical photographic representations. Certainly this bag hid the face, but it also inadvertently masked the humanity of the subject and made her/him seem even more data-like than human.

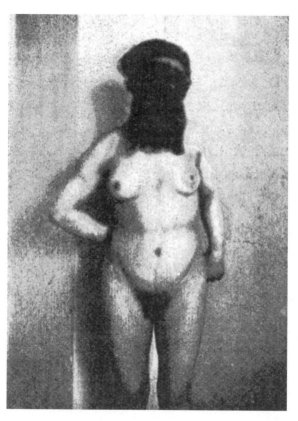

Figure 6 A hermaphrodite (probably Polish), identified only as "R.X.," with a bag over her head to protect her identity. R.X., aged twenty-one, had been raised a girl and thought herself a woman. Her fiancé had left her when she revealed to him that she could never have children. Franz Neugebauer reported to colleagues that R.X. and her father asked him to operate on her in order to make it possible for her to marry, but he told them this was impossible. The father responded by saying that he would find a rural son-in-law who would not think R.X. abnormal.

The pronoun problem may also have contributed to the objectification of hermaphrodites. Given a language with gendered pronouns—as both French and English are—it is always difficult to determine what pronoun to employ when talking about a hermaphrodite. French authors of the nineteenth century figured a way around this difficulty: they almost always referred to hermaphrodites not by their proper names, but as *le sujet* or *le malade* ("the subject" or "the patient"—literally, "the ill one"), because doing so enabled them to avoid the problem of having to pick a pronoun to match the subject's sex. These French authors simply employed the masculine pronoun because it matched the gender of the chosen noun (*sujet* or *malade*), if not the gender of the person in question. English-speakers were forced to choose, and they sometimes remarked on the awkwardness of the situation. On very rare occasions, and apparently with reluctance, they used the ungendered and unpleasant "it." On many occasions they switched from one gendered pronoun to the other. (Stuck as I am with the rigid dichotomy of English singular pronouns, I employ whichever pronoun seems to make sense in the context of my story, that is, whichever pronoun would make sense to the person whose viewpoint I am trying to present. If that is not possible, I try to use the gender by which the hermaphrodite identified him/herself.)

Despite all the objectification, however, in the texts and the photographs which record cases of hermaphroditism, one cannot help but feel the remarkable intimacy of the practitioner and patient, as hands and questions pried for the truth. All the dry clinical descriptions could not fully mask the intensity of the relationships between medical practitioner and attended hermaphrodite, relationships that were often mutually beneficial and cooperative. Battles of wills could and did erupt. A patient might refuse to be subjected to numerous, invasive examinations,[1] and a medical man might refuse to conspire to a sex he thought false.[2] Nonetheless, much of the time hermaphrodites and medical men worked together to achieve their respective aims. This chapter reviews the relevant histories of the most famous hermaphrodites and hermaphroditism experts and their interests in each other.

I intentionally juxtapose the biographies of hermaphrodites and the biographies of the medical experts who coped with them, and I do this in part because of my conviction that we need to recognize and respect the humanity of hermaphrodites. But I also provide hermaphrodites' biographies (what can be recovered of them) because we can see in them how hermaphrodites resisted objectification and simple solutions. The hermaphrodites of the nineteenth century posed very particular problems to which medical and scientific men of the time were obligated to respond. The medical treatment of hermaphroditism was therefore not simply a product of late nineteenth-century medical science and culture; it was also very much a response geared to the peculiar circumstances of contemporary hermaphrodites (or the understanding of those circumstances). In short, in some ways the hermaphrodites acted to set the parameters of the biomedical treatments.

Famous Hermaphrodites and the Reasons for Their Fame

Alexina/Abel Barbin (1838–1868) was, without a doubt, one of the most famous hermaphrodites of the nineteenth century. Many medical men referred to Barbin's case in their own reports of other hermaphrodites, as did Dandois, for instance, in his report on Sophie. Barbin's fame was not due to the particularly spectacular nature of her/his anatomy. Indeed, medical men ultimately decided the case was fairly straightforward, that Barbin was not "truly" a hermaphrodite—both male and female—but was, rather, just a "true" male who had been mistaken for a female. Medical men who discussed Barbin agreed that Alexina was properly Abel because he had testicles and other male internal organs and no ovaries. Most medical men who considered Barbin's case decided that, in the end, Barbin was "just" a hypospade, that is, a male afflicted with a deformity of the penis.[3]

Barbin's fame was, like the fame of other prominent nineteenth-century hermaphrodites, therefore attributable only in part to unusual anatomy. It was rather more due to his availability as a specimen and a spectacle. Although researchers and practitioners had started to notice patterns in the forms of hermaphroditism by the early 1800s, generally throughout the century medical and scientific study tended to focus less on particular types of hermaphroditism than on particular cases. The cases that generated the most discussion were not necessarily those that were the most unusual or representative anatomically, but those that were the most readily accessible and the most bizarre biographically. Barbin, for instance, made her/himself exquisitely available via sensational memoirs and ultimately the suicide on the rue de l'École-de-Médecine in Paris.[4]

There were plenty of hermaphrodites found and reported in the medical literature, but only a few became paradigm cases. The kind of human hermaphrodite that most excited medical and scientific men was the one who stayed alive long enough to answer pressing questions—about menstrual signs, nocturnal emissions, family history, sexual desires and acts—but who also died soon enough that the body could be located and secured, opened and preserved in drawings, photographs, wax replicas, and jars, to satisfy any observer's curiosity. Barbin's corpse and memoirs provided ample fodder for the interested masses, professional and lay. Hermaphrodites who became comparatively well known in medical professional circles tended to be those who were well documented and those discussed—or, better yet, actually examined—by medical authorities. In an interesting way, hermaphrodites and their doctors tended to amplify each other's fame.

Several medical men who examined Barbin found it unbelievable that he could have been mistaken for a girl at birth. They assumed that Barbin's parents and others who saw Alexina's anatomy must have been either ignorant of medical "facts" or inexplicably confused.[5] But those who knew the record knew that, earlier in the nineteenth century, around the time of Barbin's birth, a person named Gottlieb Göttlich (1798–?) had recounted a fantastic transformation from girl to man similar to Barbin's. Like Barbin's story, Göttlich's tale was seriously doubted by the medical men who saw him only as an adult in the 1830s. Indeed, like Barbin, the adult Göttlich probably would have been simply diagnosed a hypospadic male had he not kept exhibiting himself to more and more doctors, thereby generating confusion and controversy about his condition and his true sex—and generating profit to boot. In the 1830s Göttlich earned a respectable living, monetarily speaking, by exhibiting himself at medical schools across Europe and the British Isles.

Born on March 6, 1798, in the Saxon village of Nieder Leuba, Göttlich was presumed female and so was baptized and raised Marie Rosine Göttlich. Like dozens of hermaphrodites after her, Marie Rosine's sex was called into question when, at the age of thirty-three, an apparent double hernia drew medical attention and the suspicion that the supposedly herniated organs were in fact descended testicles. Göttlich was sent to Professor Friedrich Tiedemann at the University of Heidelberg for his opinion in November 1832. Tiedemann performed a brief examination and declared Marie Rosine was "evidently a man, with genitals of uncommon conformation. She will dress herself, therefore, in men's clothes, and adopt the name of Gottlieb."[6] Göttlich did so, and then used the combination of a male passport and an ambiguous anatomy to travel, exhibit himself, and earn fame and fortune.

It is unclear from the record whether Göttlich, Tiedemann, or someone else first suggested the possibility that Göttlich could change her legal status, but whoever initiated it, Göttlich made the most of the idea and engineered an especially positive change of social identity. With written testimonies to his masculinity in hand from the well-known Tiedemann of Heidelberg and Johann Blumenbach of Göttingen, Göttlich donned male clothing, obtained a passport that indicated his new sex, and went on the road. In just three years (1833–1835) as a traveling hermaphrodite catering to the curiosity of medical men, Göttlich voyaged to Bonn, Jena, Marburg, Mainz, Offenbach, Breslau, Bremen, Hamburg, London, Manchester, Liverpool, Cork, Dublin, Glasgow, Aberdeen, Monstrose, and Edinburgh. At medical schools, he stripped and was examined by dozens—and perhaps ultimately hundreds—of men of science and medicine who came to see this curious case and render their often-conflicting opinions as to Göttlich's "true sex." His customers included such luminaries as Blumenbach, P. D. Handyside, Sergeant-Surgeon to the Queen Sir Astley Cooper, Robert Knox, and Robert Grant. From many of the medical and scientific men who saw him, Göttlich cleverly obtained certificates which testified that his case was of deep interest to the medical man, the naturalist, the phrenologist, and the physiologist. These were used in turn to generate still more interest and profit. In fact, when a corrective operation was offered him, "Göttlich declined all surgical aid." He remained "averse to a proposal of this kind, since it would at once deprive him of his . . . easy and profitable mode of subsistence."[7] The British Library still holds an English-language copy of his "Certificates of a Very Rare Specimen of Hermaphroditism."[8]

As Robert Bogdan noted in his historical study of freak shows, many so-called freaks of the nineteenth century earned a great deal of money exhibiting themselves and were thereby able to lead financially secure lives.[9] In the United States, for instance, Chang and Eng Bunker, the original "Siamese" twins, earned enough through exhibitions to marry, buy farms, and even purchase slaves.[10] I cannot find much evidence, however, that many hermaphrodites besides Göttlich exhibited themselves for money to medical professionals or lay people in France or Britain. Göttlich's special status as a foreigner may well have enabled him to exhibit himself around the British Isles. There, and perhaps elsewhere, it seems that people of other nationalities and "races" were much more readily exhibited publicly as anatomic spectacles.

In France public exhibition by hermaphrodites may have been discouraged, as it was in the case of some conjoined ("Siamese") twins, out of a fear that if pregnant women witnessed a "monstrosity," they might pass on that monstrosity to their *in utero* children. (This concept of "maternal impressions" is discussed in more detail below.) Nonetheless, medical and scientific men jumped at the opportunity to examine a living or dead hermaphrodite. Take for instance the case of Marie-Madeleine Lefort (1799–1864). Born just as the nineteenth century was about to begin, Lefort retained the position of the most famous "female pseudohermaphrodite" from the time of her discovery until at least 1915. In fact, the story of Lefort was called upon often to remind nineteenth-century medical men faced with living people of "doubtful sex" that no matter how masculine-looking a subject, that subject could be a "true woman" by virtue of internal female organs.

Lefort died in Paris at the Hôtel-Dieu at the age of sixty-five in 1864, just three and half years before Alexina/Abel Barbin's suicide. In fact, only a few blocks away, while Barbin toiled at his railroad desk-job, Lefort had her "true" femininity—that is, her possession of female internal organs and no male internal organs—finally confirmed by autopsy. At the time of her death, Lefort was balding on top but had an even finer beard than the one she had sported at the age of sixteen when, in February of 1815, she had been presented at the Paris school of medicine and examined by several renowned medical and scientific men (see Figures 7, 8, and 9). In 1815 a commission had been appointed to examine Lefort and to determine her true sex. Two experts on the commission thought her a malformed male, but P.-A. Béclard had insisted she was really a female, albeit with fused labia majora, a hypertrophied clitoris, a strong voice, and a beard. The autopsy vindicated Béclard, who was himself dead almost forty years; Lefort's internal organs were clearly of the female type, and by then it was the internal organs that easily held sway in post-mortem matters of "true" sex.[11]

To make matters even more confusing for medical men of the nineteenth century (and for medical professionals of today), not all cases of so-called pseudohermaphroditism are as easy to spot from an external examination as Lefort's was. More and more convincing-looking "women" who were revealed to have testes surfaced as the nineteenth cen-

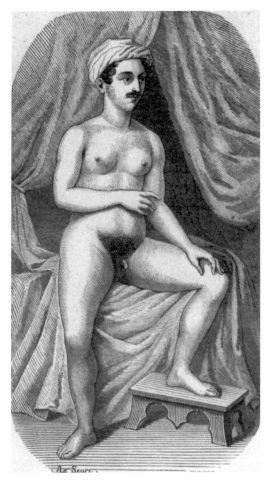

Figure 7 The famed French hermaphrodite Marie-Madeleine Lefort, aged sixteen.

tury progressed and as the scope of gynecological surgery expanded. Besides challenging the efficacy of the gonadal definition of sex, this kind of hermaphrodite also forced medical men to struggle with the idea that a "true male" could live a long life as a woman and never be suspected of being a male. They knew this had in fact happened: the posthumously famed Italian hermaphrodite Maria Arsano spent her eighty-year life as a woman, never having been suspected of being otherwise. Arsano had been married to a man for many years, and died in one of the pauper charities in Naples. Her "malformation . . . was only at first accidentally discovered in preparing the dead body for demon-

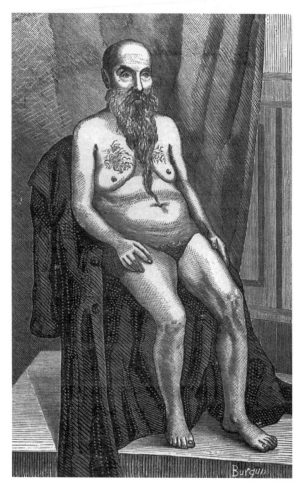

Figure 8 Marie-Madeleine Lefort at age sixty-five, shortly before her death.

stration in the anatomical theater of Professor Ricco."[12] In other words, Arsano's hermaphroditism was only discovered when she was opened after death and shown to have testes inside. (See Figure 10.) Her hermaphroditism might never have been revealed if she had died elsewhere. Variations in sexual anatomy could be very subtle and truly remarkable inside and out.

In the end, these famous hermaphrodite stories and others like them filled the medical literature on hermaphroditism and in turn filled professional readers with fascination and apprehension. Practitioners had to realize after a perusal of the literature—or even after a single personal

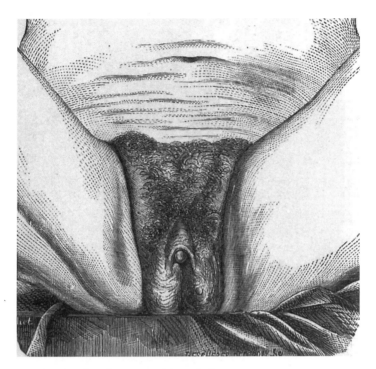

Figure 9   The genitalia of Marie-Madeleine Lefort.

encounter with a case of doubtful sex—just how messy and unpredictable sex could be.

Mutual Interests

What could a nineteenth-century hermaphrodite hope to gain from an encounter with a medical man? Most hermaphrodites sought only relief from pain, dysfunction, or nagging questions, but occasionally some sought to profit monetarily from the doubt surrounding their sex as Göttlich had overtly done by charging curious souls for a look at his ambiguous anatomy. Because the roles allotted men and women were so different, actions in cases of mistaken sex were sometimes suspected of being motivated by profit, and it was true that some hermaphrodites sought a revision of their sexual status from female to male precisely because their earning potential would immediately increase. One supposed English "woman," for instance, tried to claim she was "the rightful heir to a title" because of what she said was her "true male" sex. Her claim

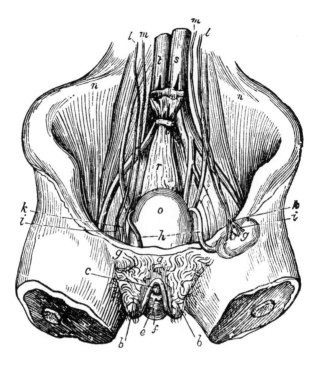

Figure 10 An anatomical sketch of the anatomy of Maria Arsano. James Young Simpson classified Arsano as a true hermaphrodite, specifically a case of "transverse hermaphroditism with the external sex organs of the female type." She would later (under Klebs's scheme) be classed a male pseudohermaphrodite. The letters indicate: a. pubic hair; b. labia majora; c. clitoris; e. urethral orifice; f. vaginal opening; g. testicles; h. vas deferens; n. ureters; o. urinary bladder; k, l, m, s, t, blood vessels.

failed, and instead the "woman" landed in an asylum, her penis-like organ suspected of actually having grown from her delusions of malehood.[13] But profit from a sex revision could be extramonetary, too, and medical practitioners recognized this. For instance, some French doctors who examined very masculine adult hermaphrodites wondered if these hermaphrodites' mothers had pretended their children were females to avoid their conscription into the army. Similarly, some hermaphrodites like Barbin thought that revisions of their sexual status might make possible relationships with otherwise forbidden lovers. The French hermaphrodite Henriette Williams, for example, was actually relieved to hear that she was a man, because she wanted "later to be married [to a woman she knew] and to modify her civil status"—presumably in the reverse order.[14] Much was at stake in matters of sexual identity, and doubt could, if played right, be very valuable to a hermaphrodite.

In sum, medical men provided hermaphrodites with diagnoses, explanations, relief, distress, confusion, wanted and unwanted sex changes, and occasionally even financial gain. Hermaphrodites gave the medical men much in return—most obviously in the form of fascinating cases for study. Indeed, medical and scientific men often found themselves drawn to cases and histories of hermaphroditism (as many of us are) by an intense curiosity that bordered on the voyeuristic. Who could blame them? The anatomies and biographies of hermaphrodites were remarkable. Sometimes they seemed unbelievable, at other times almost comical. Consider, for instance, the story of Marie-Faustine, a story often inserted in treatises on hermaphroditism apparently for a combination of comic relief and admonition about the implications of sex diagnosis. Marie-Faustine was, according to the legend, "a woman of 50, who had been married at the age of 21, and divorced from her husband ten years later. The husband's plea was not only that he could not consummate his marriage, but that his wife made him a source of ridicule to his acquaintances because of her conduct towards other women." Marie-Faustine was having sex with "other" women. "It was discovered at last that this supposed woman was in reality a man, the subject of hypospadias; and thereupon she claimed from her brother a share of the patrimony [that is, of their father's property]. Against this claim the brother entered a counter-plea against the newly discovered brother of seduction of his wife."[15]

Thus, while bearded women were sometimes exhibited as amazing and amusing curiosities in the fairs and shows of London and Paris, so too were bearded women and beardless men attractions exhibited at medical schools, at hospitals and museums, in medical texts, and at professional gatherings like the 1888 meeting of the British Gynaecological Society where Barnes had presented his "living specimen." It is easy, when reading long clinical case histories, to lose sight of the extent to which medical and scientific men found hermaphroditism an "amusing" subject, until one comes across a notice like the following one in the December 2, 1882, issue of the *British Medical Journal.* It was published as part of the short report of a meeting of the South Eastern Branch, East Kent District, of the British Medical Association:

Mr. Brian Rigden showed a [subject of] Spurious Hermaphroditism, seven months old. There was a large clitoris, not perforated, and

labia without testicles. A small opening existed beneath the clitoris. .
. . There was no visible vagina, and the finger in the rectum failed to
detect an uterus. . . . This child was considered a female by the
meeting. A very amusing and interesting discussion followed.[16]

The importance of "amusement" should not be underrated; it was not
uncommon for a medical or scientific treatise on hermaphroditism to
include simply "as curiosities" a litany of sensational stories of various
hermaphrodites' lives.[17] These stories were told and retold in medical and
scientific texts, and inevitably they were sometimes embellished. Obviously curiosity and a sort of voyeurism were part of what drew interest to
the subject of hermaphroditism.

Professional Reputations

Still, amusement was only a small part of what medical and scientific men
gained from their encounters with hermaphrodites. The former also stood
to gain professionally from their intellectual and clinical treatment of
hermaphrodites. Individual cases of hermaphroditism offered a medical
man a chance at professional renown, or professional infamy if his handling of a case was considered exceptionally poor. Hermaphrodite studies
were a lively storytelling genre in medicine, and if a case came with a good
story, chances were that the case and by consequence the medical man
who described it would not be forgotten when the litany of hermaphrodite lore was recounted along with the details of a new case. If a medical
man could preserve a hermaphrodite's organs for additional study, so
much the better for the preservation of his name. Probably in part out of
a concern for patients' privacy and in part as a result of an emerging
professional identity, as the nineteenth century progressed, famous cases
were more and more likely to be named after their dominant investigators
instead of the hermaphrodites themselves. For instance, the "case of
Marie-Madeleine Lefort" slowly became known as "the case of Béclard."
   Medical men also stood to gain the trust and respect of those who came
to them with cases of doubtful sex. In Britain and France, medical men
were increasingly consulted by parents of doubtful-sex children, although
sometimes parents hid their malformed offspring out of shame or relied
on a midwife or trusted family member for advice.[18] Sometimes medical
men seem to have deferred to their clients' wishes against their better

judgment with regard to sexual diagnosis—although this seems to have been the case almost exclusively when the client was of an elevated social status.[19] Of course, the general issue of the social status of medical practitioners was at stake in these extraordinary encounters. The late nineteenth century was an important period of professionalization for medical men in France and Britain, and we can see in their encounters with hermaphrodites a striving to become trustworthy and vital authorities in important social matters. Specifically, in matters of doubtful sex, medical men hoped to offer the "direction of a skilled opinion" and to thereby become the "authorized interpreters" of "true" sex and sexual identity.[20] By consulting with medical men, hermaphrodites and their parents supplied an important acknowledgment of the medical men's growing authority, and confirmation that the medical men were indeed the just and trustworthy arbiters of pathology and identity. In these encounters, then, there occurred a simultaneous negotiation of the identity of both groups, of the medical and scientific men, and of the hermaphrodites themselves.

As interest in and concern over hermaphroditism grew in the late nineteenth century, a few medical practitioners went beyond association with a single hermaphrodite to become experts in the field. The most prominent of these was neither French nor British, but Polish. He figures importantly in this story, since as an active member of Polish, German, British, and French medical societies, he disseminated information and advice on the subject, advice upon which French and British medical men relied. This man was Franciszek Neugebauer (1856–1914; see Figure 11).[21]

Following in the footsteps of his proto-gynecologist father Ludwik Neugebauer (gynecology was just emerging as a specialty), Franciszek Neugebauer decided to pursue a medical degree and become a gynecologist. He studied first in Warsaw and then in Dorpat, where he earned a doctorate in medicine. Before returning to Warsaw in 1887 to assist at his father's gynecological clinic, Neugebauer furthered his education in Leipzig, Dresden, Berlin, Paris, London, and Vienna.[22] His command of languages was impressive. Besides his native Polish, Neugebauer could speak and write fluently in French, German, and English.

In 1897 Neugebauer gained the appointment of chief of the Gynecological Department at the Evangelical Hospital in Warsaw, a position he would hold for the remainder of his life. From his Warsaw clinical base Neugebauer developed a reputation as a specialist in two areas, hermaphroditism and inborn deformities of the pelvis. Subjects of these malformations were sent to him from all parts of Poland and beyond, and

Figure 11 Franciszek (Franz) Neugebauer (1856–1914), expert in the field of hermaphroditism.

Neugebauer himself traveled to investigate reports of other cases in many nations. He carefully documented the details of cases in writing, photographs, and casts, and he readily shared his collections with other practitioners.[23]

All told, a sizable portion of Neugebauer's time and energy was devoted to his journeys to meet fellow practitioners and educate them regarding his findings. He made a regular practice of attending meetings of medical associations in Europe and the United States, and he was elected to the membership of many medical and scientific societies, including the British Gynaecological Society, the Warsaw Medical Association, the Warsaw Science Association, the New York Obstetrical Society, the Paris Academy of Medicine, the Association of Pathologists in Brussels, and the Belgian Royal Academy of Science.[24]

Neugebauer's notices on hermaphroditism consisted mostly of large collections of observations,[25] and clearly Neugebauer was at heart more a collector than a theorist. His notices were published in Poland, Germany, France, and Britain, and they culminated in the publication in 1908 of his magnum opus, *Hermaphroditismus beim Menschen* (Hermaphroditism in

Man). This massive tome included summaries of some two thousand cases of human hermaphroditism recorded since the beginning of history from around the world. Neugebauer had observed roughly forty of these cases first hand.

France also boasted a hermaphrodite expert by the turn of the century, the Parisian surgeon Jean Samuel Pozzi (1846–1918; see Figure 12). Samuel Pozzi was born in southwestern France, the son of a Reformed Church clergyman.[26] He became a student of medicine in 1869 at Paris where he studied under Paul Broca, the master of physical anthropology.[27] In 1873 Pozzi received his M.D., and thereafter worked as a staff surgeon for a number of Parisian hospitals. By 1875 he had acquired an associate position with the Paris Faculté de Médecine, and in 1901 a chair of clinical gynecology was created there especially for him.[28]

Figure 12 Jean Samuel Pozzi (1846–1918), surgeon and expert in the field of hermaphroditism.

A man of "world-wide reputation," Pozzi authored or coauthored a number of important surgical and gynecological monographs, many of which were translated into English.[29] Members of the French and British gynecological communities often relied on his well-illustrated textbooks for information and advice.[30] In 1897 Pozzi founded the widely respected journal *Revue de gynécologie et de chirurgie abdominale.*[31] Neugebauer used this journal as an outlet for his hermaphrodite collections. Although particularly interested in gynecological surgery, Pozzi performed many other types of abdominal surgery, and it would be anachronistic to classify this Frenchman narrowly as a gynecologist.[32]

The widespread establishment of hospital clinics in the nineteenth century made it possible for medical practitioners like Neugebauer and Pozzi to observe numerous cases of relatively rare conditions such as hermaphroditism. In his career, Samuel Pozzi personally examined no fewer than nine living hermaphrodites.[33] He frequently published notices of those examinations and his hypotheses relating to hermaphroditism.[34] Throughout his career Pozzi concerned himself especially with the diagnosis and surgical treatment of abdominal tumors in hermaphroditic and nonhermaphroditic patients. Pozzi also took a keen interest in embryology, and used the opportunities presented by hermaphrodites' unusual anatomies to discover embryological homologies in males and females.

Like Neugebauer, Pozzi frequently traveled to confer with colleagues in Germany, England, Italy, Austria, and the United States. Contemporaries recalled that, "gowned in a white overall and wearing a black cap of Florentine fashion, [Pozzi] was a picturesque figure as he toured his wards with assistants and students . . . A man of natural charm, he was specially attentive to the foreigner who wished to follow his clinical practice."[35] Many foreigners and countrymen did. Pozzi was considered one of the most innovative, dedicated, and talented surgeons of his day.

Pozzi's career constitutes a microcosm of the rise of the professional, hospital-based, medical doctor-surgeon in late nineteenth-century France. His long career and life came to a tragic end in June of 1918. By the time of Pozzi's death, an obituary of him lamented, it seemed "as though doctors were attaining to kingly and presidential distinction," for like a king or a president, Pozzi had been "assassinated." Still practicing in 1918 at the age of seventy-two, Pozzi was confronted in his clinic by an apparently deranged patient who fatally shot the surgeon before killing himself.[36] Pozzi's observations on hermaphrodites long outlived him, particularly his latter-day, therapeutically important observation that undescended testicles in male pseudohermaphrodites often become cancerous.

No hermaphrodite specialist with the level of expertise displayed by Neugebauer and Pozzi emerged in Britain from the 1870s to the 1910s. Before and throughout this period, Sir James Young Simpson's lengthy 1839 article on hermaphroditism, discussed in Chapter 5, prevailed as the primary reference for British practitioners. In Britain in the late nineteenth century, the task of providing up-to-date advice in cases of doubtful sex fell, if to any particular British medical practitioners, to Robert Lawson Tait (1845–1899) and Robert Sydenham Fancourt Barnes (1849–1908). Still, it was just as likely that a British practitioner faced with a case of doubtful sex would turn for guidance to the latest writings of Neugebauer or Pozzi, or to the comparatively outdated article by Simpson.

Fancourt Barnes was the eldest son of Robert Barnes (1817–1907), the renowned obstetrical physician, but Fancourt never attained the prominence or productivity of his remarkable father. Indeed, the younger Barnes was said to have ridden on the coattails of his father,[37] a practice not necessarily beneficial given the contentious relationship the elder Barnes maintained with many of his colleagues during Fancourt's formative professional years. Fancourt "made his mark early," in the 1880s and early 1890s, and then retired from professional life because of problems with his health.[38]

After a good deal of professional strife, in 1884 the two Barneses, along with Neugebauer and Tait, broke from the Obstetrical Society of London to form the British Gynaecological Society.[39] This society was the main vehicle of Fancourt Barnes's work, and it was at one of its 1888 meetings that Fancourt made his presentation of a "living specimen of a hermaphrodite." Beyond that presentation, Barnes published a small number of notices of hermaphrodites.[40] With his father he also provided a basic discussion of hermaphroditism in their 1884–1885 textbook, *A System of Obstetric Medicine and Surgery.*[41]

A notably more colorful figure, the gynecological surgeon Lawson Tait was "no ordinary man."[42] Born in Edinburgh to Isabella Stewart Lawson and Archibald Campbell Tait, who was a guild brother of Heriot's Hospital, as a boy Tait was educated at Heriot's Hospital school. He then studied medicine at the University of Edinburgh but he never finished his degree, apparently out of impatience. He eventually received the M.D. only as an honorary degree, from the State University of New York.[43]

Much of Tait's early education came from James Young Simpson, author of the article on hermaphroditism already mentioned. Tait studied under Simpson and assisted the elder man in his obstetrical practice.[44]

A fast-rising star, in 1866 Tait was made a licentiate of the Royal College of Physicians and the Royal College of Surgeons of Edinburgh, and he became a member of the latter and a Fellow of the Royal College of Surgeons of England in 1870 and 1871 respectively. Tait spent most of his later years in Birmingham, where he "took an active interest in the public life of the locality" including in matters of politics, art, and public health.[45] One of his most durable contributions to surgery was his advocacy of aseptic practices, which he championed over Lister's antiseptic methods.[46] Lawson Tait was a memorable figure, known for "his clear-headedness, his pugnacity, his contempt for detail, his worship of good surgery, and his glory in his own talents."[47] Contemporaries recalled that he cast a memorable image with "his short burly figure, his leonine head, his determined mouth and his masterful expression."[48] In the medical community of the time, Tait was perhaps best known for his strong advocacy of ovariotomy (removal of the ovaries), a procedure he himself performed hundreds of times, believing as he did that such a procedure would cure the patient of a whole host of problems, physiological and behavioral.

Throughout his life Lawson Tait exhibited an avid interest in natural history, biology, and histology, and his writings and lectures were frequently littered with enthusiastic endorsements of Charles Darwin's theory of evolution.[49] Indeed, Tait explained human hermaphroditism with Darwinian theories of evolution, and he in turn used instances of human hermaphroditism to bolster Darwin's theory. He insisted:

> We must accept Darwin's theory of the descent of man. This acceptance at once becomes the explanation of the occasional occurrence of bisexual vertebrates, and consequently of true hermaphroditism in human individuals. Conversely, the occurrence of such malformations may be offered as one amongst the many proofs which are being accumulated from every quarter in favor of Darwin's theory.[50]

Many of Tait's colleagues found this argument compelling.[51] It seemed logical to them that the greater frequency with which the inferior vertebrates produced "monstrous" hermaphrodites spoke both to their proximity to a common hermaphroditic progenitor and to the existence of that progenitor.[52] (Darwin thought such a concept correct, but also amusingly scandalous. He kidded his friend, the geologist Charles Lyell, "*Our* ancestor was an animal which breathed water, had a swim bladder, a great swimming tail, an imperfect skull, and undoubtedly was an hermaphrodite! Here is a pleasant genealogy for mankind!")[53]

Tait himself wrote just one short treatise dedicated exclusively to the subject of human hermaphroditism, and that was published only in the United States.[54] He did, however, observe a small number of hermaphrodites in the course of his practice and, like Neugebauer and Pozzi, apparently had a few hermaphroditic subjects referred to him.[55] Tait's contributions to the field of hermaphrodite studies came mainly through brief discussions of it in his general texts on gynecology (for example, *Diseases of Women,* 1879) and through his joint investigations into cases of hermaphroditism, like that at the April 1888 meeting of the British Gynaecological Society. Tait's reputation among the British as an authority on doubtful sex appears to have been a derivative of his reputation as a leading gynecologist.

Gains for Science and Medicine

Participation in the treatment of hermaphroditism certainly contributed to the professional reputations of men like Neugebauer, Pozzi, Barnes, and Tait, but the interest of these men was by no means exclusively self-serving. These men were also clearly motivated by a shared desire to improve scientific knowledge and medical practice. The ever-growing corpus of clinical records of hermaphroditism aided many British and French physicians and surgeons in diagnosis and treatment. For instance, a well-read practitioner of the late nineteenth century knew that the record showed that most doubtful people were "true males" afflicted with hypospadias, and so he knew that in a doubtful case his best guess was "male."[56] Similarly, surgeons could make good guesses about hermaphrodites' likely anatomy and about where to cut by noting the patterns of genital malformations documented in the medical literature.[57]

Instances of hermaphroditism also promised to shed light on many realms of scientific study, including embryology, sex determination, evolution, and questions about the role of internal secretions (later called hormones) in human physiology. Indeed, hermaphroditic beings offered material for study in a very literal sense, in the form of organs: the nineteenth century saw a minor trade in reproductions and parts of hermaphrodites. Medical and scientific men purchased nonhuman hermaphroditic specimens from fishmongers, from sheep farmers, and from butchers,[58] and they also lent and sold each other wax casts, drawings, photographs, and preserved specimens taken from human hermaphrodites.[59] "Valuable" specimens were sometimes literally so.

Medical and scientific men hoped and believed that a thorough investigation of the phenomenon of hermaphroditism would allow them to rein it in theoretically and practically. Among their chief goals in the study of hermaphroditism were first, a general reduction of the disturbing uncertainty that surrounded the phenomenon of human hermaphroditism—a reduction that would come from applications of the scientific explanation for its origin; and, second, an elucidation of the details of the developmental relationship of the "normal" female and "normal" male, again, in part so that they might possibly prevent cases of human hermaphroditism. Medical and scientific agendas in cases of hermaphroditism could never be categorized as purely intellectual or purely practical. Theory and practice developed together intentionally. This was in keeping with the modern tradition of teratology. In hermaphrodite studies as in studies of other "monstrosities," exploration, explanation, alleviation, and prevention formed joint pursuits.[60]

Embryology

How then did medical and scientific men of the nineteenth century try to explain hermaphroditism? As we saw in the last chapter, the dominant motif of scientific discourse regarding human males and females held that, at least externally, the male presented a more elaborately developed form than the female; the female represented a lesser degree of development. In general the male appeared to progress several steps beyond the female in genital development. It seemed logical therefore that arrests—haltings or delays—in development could make a male look like a female, while excesses in development could make a female look like a male.[61] (In nineteenth-century teratology, arrests and excesses of development were used to explain all sorts of malformations, not just those resulting in sexual ambiguity.)[62] Not atypically, "for clinical purposes," Lawson Tait divided hermaphrodites generally along these developmental lines into two basic groups: "those in which an arrest of development in the male organs give them an appearance as if the child belonged to the female sex," so-called masculine pseudohermaphroditisms; and "those in which an excessive development makes the female organs resemble those of the male," so-called feminine pseudohermaphroditisms.[63] Explanations for particular hermaphroditic cases and types almost always posited arrests of development in males or excesses of development in females.[64] For instance, the "arrested" development of hypospadiac males was often compared to the normal development of females, and Pozzi even pondered whether "woman, from the point of view of the external genital organs, is not a veritable hypospade."[65]

This internal/external organ distinction made by Pozzi is important, for he and his colleagues knew that, with regard to the internal organs, the female type involved not a mere underdevelopment of the male type, but a clear divergence in development. In the female fetus, the two proto-gonads took one path to become ovaries, while in the male fetus, they followed another and became testes, as noted in Chapter 1. And, while ultimately-male and ultimately-female fetuses both began with Mullerian and Wolffian systems of proto-organs, in the female, the Wolffian system atrophied and the Mullerian system evolved to form "female" internal organs, whereas in the male the Mullerian system atrophied and the Wolffian system evolved to form "male" internal organs.

These details of internal sex differentiation were often recounted in texts on hermaphroditism as part of explanations for anomalous internal sex organs.[66] A "true male," for instance—that is, a person with testes—could appear to have fallopian tubes if for some reason the Mullerian system had not atrophied in him as is typical in males. Inversely, such "teratological variations" were used as clues to the embryonic origin of certain structures of sexual anatomy.[67] Pozzi, for instance, performed an adroit comparative study of several different kinds of hermaphroditisms to demonstrate that the female hymen derives from an embryonic origin common with the external vulva, and not with the internal superior vagina.[68]

Hermaphroditic humans were also used as living proof of "the primitive bisexuality and the primordial hermaphroditism of the embryo."[69] Even while the female was often portrayed rhetorically as a sort of under-evolved male, medical and scientific men thus professed the conviction that the male and female types actually diverged from a common, original hermaphroditic state in both embryology and evolutionary history. The British surgeon Jonathan Hutchinson, for instance, suggested that "like many other conditions, hermaphroditism is a thing of degree, and . . . up to a certain point all persons are bisexual." He reminded his readers:

Up to a certain age the foetus has potentially the organs of both sexes, and it is only by the ascendant development of the one set that the other is suppressed. Nor is the suppressal [*sic*] in either sex ever absolute, for every male has mammary glands and every female has a clitoris, organs which definitely belong to the other sex, and which persist only because the suppressal has been incomplete.[70]

Teratological postembryonic human hermaphrodites were therefore to be understood to be "individuals in whom a part of the organogenic scaffolding of the genital organs, which is common to both sexes, persists instead of atrophying."[71]

Maternal Impressions and Hereditary Antecedents

Embryology helped explain how human hermaphroditism could occur, but it seemed to offer little hope of explaining why it did occur. No doubt from a combination of understandable curiosity and a hope that human hermaphroditism could be prevented, medical men never abandoned their profound interest in why genital malformations resulted from certain pregnancies and not from others. On this subject, many experts chose to remind colleagues that "a psychic influence on the part of the mother, and the factor of heredity, cannot be excluded in face of the instances in which several members of the same family . . . have suffered from the same malformation of the genital organs."[72] Indeed these two factors, maternal impressions—that is, thoughts of the mother literally impressed upon and embodied by the developing child—and hereditary antecedents—that is, faults of the mothers and fathers—were the factors most often blamed for human hermaphroditism throughout the subject period.

British medical men were more likely than their French counterparts to seek out and relate evidence of maternal impressions in cases of hermaphroditism, although Pozzi certainly did not doubt that "the influence of strong emotions on the production of monstrosities seems established by very solid proofs."[73] Most expected that an emotional or sometimes physical trauma would have struck a hermaphrodite's mother in the second or third month of the pregnancy, that trauma thereby forcing sexual development off its normal path.[74] Fancourt Barnes, for instance, was not surprised to hear that the mother of his "living specimen" had "had a fright when about two months pregnant."[75] Pozzi similarly recorded with regard to the mother of another hermaphrodite that "when she was pregnant with [the hermaphrodite], at three months, she had had a great fright; a man was crushed in her presence."[76]

Interestingly, although hermaphrodites' mothers were often considered by medical men poor witnesses with regard to observations of the "sex" behavior of their doubtful children, they were frequently expected to account for the deformity in their offspring.[77] And often enough, the mothers did confess some fright, fear, or wish during pregnancy that they now worried might have caused their children's malformation. In Britain, several medical men thought it possible that the presence of more than one hermaphroditic child in a family might be due to a sort of maternal-impression inertia. One English mother considered "it possible that the causation of the deformity [of her children] may have been in the case of the elder child a great desire on her part that it should be a girl, and in the case of the younger, brooding over the deformity of the elder."[78] The Scottish mother of three hermaphroditic children discovered in 1893 had a similar tale of self-fulfilling prophesies of doom:

> Previous to the birth of the first of these three [hermaphroditic children] they lost two children by scarlet fever, and the mother suffered a miscarriage. She was much distressed in mind at this time [by the loss of her other children], and she attaches importance to this point. She says, too, that after the birth of Robert and Jessie [both hermaphroditic] respectively she brooded much on their defects [while pregnant with Lizzie who also turned out to be hermaphroditic].[79]

It seemed a mother's fears and desires could be made flesh in her child.

British medical men also considered it very likely that certain forms of hermaphroditism, including some types of hypospadias, would have a hereditary component to them,[80] and so they quizzed hermaphrodites and their mothers about the condition of other members of the extended family.[81] Some even went so far as to contact other medical doctors to ask them to track down, examine, and watch siblings, cousins, aunts, and uncles.[82] By 1909, swept up in the belief in the power of heredity, a member of the Galton Eugenics Laboratory at the University of London assembled pedigrees of families of hermaphrodites in an attempt to discern hereditary patterns.[83]

In the 1860s and later, Charles Darwin showered his compatriots with mounds of evidence that "monstrosities" could arise from consanguinity, and indeed in Britain hermaphroditism was occasionally explained by reference to intermarriage. For example, the mother of a pair of hermaphrodites,

> a particularly intelligent woman . . . of her own accord mentioned [to a medical doctor] that she and her husband were second cousins, and she was strongly of opinion that this consanguinity was the cause of her having given birth to the two hermaphrodites. In proof of her contention she told [the medical doctor involved] that her husband, who was a market gardener, had noticed that if he planted seeds taken from cabbages which had grown near each other the resulting plants were not at all as vigorous as would be the case when he took the seeds from cabbages which had been grown at a distance from each other.[84]

The doctor who saw the younger hermaphrodite child was inclined to agree with the mother's theory.

In France, meanwhile, where generalized theories of degeneracy were used to explain many sorts of physical and moral "perversions," the search for "hereditary antecedents" included inquiry into all types of shortcomings of the hermaphrodites' parents and other relatives. Were the parents consanguineous, or afflicted by "mental alienation"? Were other members of the family malformed? Were they melancholic, or prone to drinking too much?[85] All these factors could count as sufficient causes for arrests and excesses of development leading to hermaphroditism. Pseudohermaphroditism in France constituted "a grave stigmata of hereditary degeneracy,"[86] and one of many reasons to avoid and prevent—even prohibit—consanguineous marriages, drunkenness, and a host of other social ills.[87]

Direct Material Causes and Endocrinology

In matters of hermaphroditism, most practitioners and theorists in France and Britain posited causal factors like the ones just named for which the etiological link remained unarticulated, even obscure. A few, however, did suggest some possible direct physical causes. One French theorist, Jean Tapie, wondered whether, if a teratogenic "embryo-amniotic [sac] adherence" occurred, could it not happen that development would be delayed or arrested so that a truly male child could have the sort of feminine appearance his particular hermaphroditic subject did? Tapie thought it very likely, especially given his own recent research into hand malformations and the similar findings of Isidore Geoffroy Saint-Hilaire, and of Haller and Meckel before him.[88] Geoffroy's teratology left a strong legacy in France, and so French medical men sometimes suggested proximate causes for genital malformations which harkened back to the so-called anatomical philosophy and the speculative theories of Geoffroy.[89] But most agreed that the direct developmental cause or causes of genital malformations remained unknown.[90]

Given the remarkable findings of research into internal secretions (what would later come to be called "hormones") in the late nineteenth century in France and England,[91] and given the keen focus placed on the gonads in actual cases of hermaphroditism, one might expect that endocrinological (hormonal) explanations for hermaphroditism would have emerged early in the twentieth century. Indeed, from about 1904 on, medical and scientific reporters of cases of hermaphroditism occasionally offered an explanation of their subjects' conditions that centered on the physiology of the gonads. However, it was not until after the publication in 1916 of Frank R. Lillie's investigations into free-martin (hermaphroditic) cattle that there emerged any organized, generalized theory of the possible role of hormones in the production of hermaphroditism.[92]

Two basic sorts of endocrinological hypotheses of the etiology of hermaphroditism did start to materialize before 1916. The first was that the presence of an ovotestis, that is, a sex gland containing both ovarian and testicular tissue, could explain, via internal secretions, a mix or succession of male and female secondary characteristics in a single human being.[93] In 1904 two French anatomists presented their findings from the dissection of a goat which had ovotestes. They suggested that perhaps "in the Mammals, the physiological and anatomical manifestations of hermaphroditism are due to the combined action, most often in unequal proportions, of the two interstitial glands, the one male and the other female, which are abnormally developed side by side in the genital glands."[94] S. G. Shattock and C. G. Seligmann offered a similar hypothesis in 1906 to the Pathological Society of London.[95] Their subject was a two-year-old Leghorn fowl who possessed ovotestes and an odd mix of secondary sex traits. Shattock and Seligmann concluded from their study that "the ultimate criterion of sex resolves itself into the structure of the reproductive glands."[96] More specifically this meant that "the transformation of plumage in the female of certain birds to that of the male, accompanied as it sometimes is by the cessation of laying and the acquirement of male instincts, indicates an hermaphroditic condition of the sexual gland, the *male* constituent of which commences to function after the atrophy of the *female*."[97]

But such findings were often contradicted by a close analysis of cases of alleged ovotestes in humans, in which, generally, the subjects seemed to their observers to be predominantly feminine.[98] In humans with supposed ovotestes, it seemed generally that either the testicular portion of the gland was not functioning, or there was not a simple one-to-one mapping of sex-gland tissue to secondary sex characteristics. Moreover, most hermaphrodites simply did not seem to have ovotestes or both testicles and ovaries. This explanation could not account for all, or even most, of what observers saw.

The second emerging hypothesis of internal secretions and hermaphroditisms posited that a failure of the testicles to descend might result in an arrest of development in a male, with the result that he would look, and perhaps act, more like a female. In 1899, for instance, in France, Delagénière suggested that a failure of testicles to descend might prevent the onset of masculine secondary sex characteristics.[99] In England in 1909, Samuel Wilks wrote to the *British Medical Journal* in order to publicize his conviction that most cases of so-called hermaphroditism were really instances of undescended testicles. This thesis was "corroborated in the cases where the testicles are removed soon after birth. In the human subject the appearance during growth becomes more feminine . . . like that of a woman."[100] Wilks also appealed to comparative anatomy and physiology to make his case, citing Gilbert White's observations on the emasculating effects of castration in "man, beast, and bird."[101]

The problem with Wilks's suggestion was that there were plenty of documented cases of very womanly hermaphrodites whose testicles had indeed descended.[102] Wilks's theory also could not explain why some hermaphrodites with ovaries looked so very manly. Endocrinological considerations of hermaphrodites seemed generally to confuse the study of internal secretions more than help it.[103] This may be why so few medical and scientific men of the nineteenth century attempted to explain hermaphroditism via internal secretion theories.

Nonetheless, as the twentieth century progressed, Neugebauer and Pozzi began to make some sense of things endocrinologically when they took note of how often the ovaries, testes, and adrenals seemed to be aplastic in pseudohermaphrodites.[104] In 1912 J. Hammond at the School of Agriculture in Cambridge suggested, after dissecting a hermaphroditic pig, that perhaps "during development the gonad has an internal secretion which controls the growth of the accessory organs." This would mean that "should the growth and development of the gonad be prevented, then the development of the accessory organs will be upset."[105] Histological evidence from the glands of hermaphrodites seemed to suggest this possibility and, moreover, that perhaps the sex glands were not the only ductless glands responsible for sex characteristics, or responsible via malfunction for hermaphroditism.[106] But only later in the twentieth century would hormones become the proximate cause of choice.

The Many Social Threats of the Hermaphrodite

Undoubtedly because of the near-constant international exchange of ideas and reports, medical discourse on hermaphrodites in the late nineteenth and early twentieth centuries tended to follow a standard pattern, no matter what the nationality or disciplinary affiliations of the author. The typical report opened with a first-person account of how the reporter came upon the hermaphroditic subject. There would follow a description of the hermaphrodite's characteristics—physical, behavioral, and biographical—in varying detail. Often these texts also contained background information on the types or likely origin of hermaphroditism (and, as already noted, a few amusing anecdotes). The implied "morals" of these narratives usually included some or all of the following: that the anatomy is the locus of truth; that the doubtful patient really had a single true male or female sex; that men and women were fundamentally different and that they should (and would) be true to their sexed natures; that if they were not, bad things would happen; that the medical man was the appropriate arbiter of truth; and that the medical man must do what he could, with theory and practice, to solve hermaphroditism.

It seemed that scientific study might help in the prevention and perhaps even the cure of hermaphroditism. But in the meantime, it was up to the medical practitioner to uncover any potential case of "mistaken" sex as soon as possible. There was plenty of reason to do so. Medical men were painfully aware that Barbin had seduced his virginal woman friend, Sara, and that this seduction had been made possible by early incorrect diagnosis of Barbin's sex and his consequent masked entry into sanctuaries of women—the convent, the boarding school, and finally the lady's bedroom. The lurid details of the affair spelled out so graphically in Barbin's memoirs became widely known after 1874 through Tardieu's publication of the autobiography, and readers could not but reach the conclusion that Sara's sanctity might have been saved had a medical man intervened sooner and forced the rectification of Barbin's "true" and social sex.[107]

But seduction was not the only form of sexual chaos at issue; mistaken sex could also lead to accidental "homosexuality." By 1903 Neugebauer could cite no less than sixty-eight "marriages between persons of the same sex, in consequence of such erroneous declarations!" In a 1903 communication to the British Gynaecological Society, Neugebauer bemoaned "what disastrous consequences may follow an erroneous declaration of sex, and that not merely for the individuals immediately concerned, their family and connections, but for others besides."[108] Indeed, Neugebauer and many of his colleagues were convinced that the social disorder which resulted from misdiagnosed sex ran so deep that it extended well beyond the immediate problem of same-sex relations and scandalous seductions. It seemed that a suppressed true sex would manifest itself in all sorts of socially disruptive behavior.[109] One paradigmatic case will illustrate this fundamental assumption.

On February 15, 1888, Dr. Jean Tapie, professor at the School of Medicine in Toulouse, was invited by his "excellent friend" Dr. Toujon to visit a young girl aged about fourteen years.[110] Toujon had been summoned by the child's mother to treat an abscess on the girl's leg, but the mother also had another concern: "her daughter did not appear to her normally formed."[111] This latter news prompted Toujon to call on Tapie, because he knew Tapie maintained a strong interest in "monstrosities."[112] It was not hard to locate the child for an examination, because she had been imprisoned for having committed a number of thefts. Her mother confessed with frustration that the girl was a thorough-going truant. The child did not lack talent for embroidery, but preferred spending her time reading trashy novels or playing rough games with "the young boys who run in the street." She had no taste for the companionship of other girls, and seemed to make endless trouble. The child was thus thought to be afflicted not only with genital deformity but with "a true moral perversion" as well.[113]

A complete examination led Tapie to conclude that the concerned mother's daughter was in fact a son, specifically "a very profound hypospade, with a bifid scrotum."[114] The chest was flat, with "not a trace of mammary glands."[115] A small penis (about three centimeters long) was found, stretched backward and bound down. The urethra opened, not at what would be considered the tip of the malformed phallus, but at its base. The apparent labia "really" constituted a bifid scrotum, of which each half contained a testicle, the central proof of the subject's sex.[116] Given the extent to which this boy's genitalia looked like the feminine type, Tapie could imagine why an "error of sex" had been committed when the child's sex had been declared shortly after birth. "Nonetheless," the professor insisted, "an attentive examination would have revealed the absence of a veritable vulva and of a vagina, and that fact alone would have been enough to signal and to avoid a mistake which never goes without grave consequences."[117]

Tapie saw evidence in this case of many such "grave consequences," for he was convinced that the child's truancy arose from the conflict between his natural sex (male) and the mistaken sex (female) according to which he had been forced to live:

> Torn between the instincts of his [true] sex and the social exigencies of the sex of which he wears the clothes, the child has been subject, in his intellectual and moral development, to a torture, a perpetual constraint, and from this wrestling . . . between the natural movements and the demands of education, there can result a true moral perversion.[118]

In short, the child's disobedience was assumed to be a result and a sign of the true sex trying to overcome the "torture" of a mistaken sex. Tapie expressed

> the conviction that when the civil status of this child is rectified, when he is rendered to the occupations of his age and of his [true] sex, he will gain a moral rectitude, and we will have put right a being who, if the error had not been recognized, would certainly [continue to] be devoted to misfortune.[119]

The medical man would, it seemed, rescue dozens from harm with a single discovery of a masked "true" sex. Indeed—even putting aside all of the many scientific benefits—what greater incentive did he need than this to involve himself with the problems of doubtful sex? Hermaphrodites gave medical men plenty to do.

CHAPTER 3

In Search of the Veritable Vulva

I'm old-fashioned. I like two sexes.

—Spencer Tracy in "Adam's Rib"

IN 1903 THE HERMAPHRODITE EXPERT from Poland, Franz Neugebauer, asked the British Gynaecological Society, of which he was a founding member, to publish in its journal an important and lengthy
communication from him. The piece, entitled "Hermaphrodism in the
Daily Practice of Medicine; Being Information upon Hermaphrodism
Indispensable to the Practitioner," constituted only one part of Neugebauer's grand mission of documenting, publicizing, explaining, and alle-
viating cases of ambiguous sex. Even from the article's title we can see that
by 1903, Neugebauer and the society felt that hermaphroditism was a
common enough problem that nearly all practitioners engaged in "the
daily practice of medicine" needed an education in hermaphroditism.

Neugebauer also believed that, in addition to being a fairly regular
problem, hermaphroditism was a particularly serious problem. He named
as a principal aim of his communication the desire to impress upon his
British colleagues "the grave consequences of an erroneous determination
of sex, and the frequency of such mistakes in ordinary practice."[1] Similar
articles were published by Neugebauer in the French medical press, and
perhaps British and French practitioners of the time needed to be admonished that "[t]he number of cases in which the surgeon is confronted by
pseudohermaphrodism is much larger than is generally supposed."[2] Even

if Neugebauer's colleagues had to be put on notice about the frequency of "mistaken" sex, however, they hardly needed reminding of the tremendous gravity of it. By 1903 they knew very well from the medical record "the importance of making early discovery of such cases, in order to save the miserable consequences, unhappiness, divorce suits, even suicides, which may follow if they are not recognised and are allowed to proceed in error."[3] Neugebauer's forever-mounting collections of "accidental" marriages (like Sophie V.'s) between persons discovered to be the "same sex" also made frighteningly clear the magnitude of "the responsibility which is incurred by a doctor when giving a definite opinion in cases of doubtful sex."[4]

"Mistaken" Sex and Consensus-Based Sex

As time went on, more and more medical men reported what they thought to be incidences of patently "mistaken" sex, that is, cases in which a patient's "true sex" was able to be diagnosed, but for some reason had been incorrectly labeled in the past. A number of medical men who encountered hermaphrodites discovered that a quick peek in the trousers or under the skirts was sometimes sufficient to reveal a case of obviously mistaken sex (or so they thought). In a good number of instances it seemed the genitalia were not really "ambiguous" at all, but just had been a little odd, and therefore mislabeled, at birth. (Barnes had at first thought his 1888 case such a situation.) Socially and psychologically the resolution of these cases of "mistaken sex" could be rather sticky, but medically it seemed a cinch.

How could such "mistakes" continue into adulthood? Secrecy about or unfamiliarity with genital anatomy sometimes allowed sex-identification "errors" to remain unnoticed even long after pubescent sexual blooms burst forth. J. Halliday Croom, M.D., was presented with "two supposed girls, aged respectively 19 and 21 . . . brought . . . by their mother with simply the history that they were both quite well but that they had never menstruated." Croom noted that these two had "masculine features," flat and somewhat hairy chests, "loud and harsh" voices, and that "the elder had a budding moustache."[5] With what he thought a stunning preponderance of masculinity, Croom had little trouble diagnosing the problem, one that the mother had apparently not suspected: the "sisters" were "really" malformed brothers. "Each had a representative of the penis, about 1 inch long," and "no vaginal orifice could be found." Moreover, "it was made out that each labium contained what was evidently a small testicle."[6]

Few medical men shown the evidence Croom described would have disagreed with his diagnosis, though many may well have wondered how it was that "such a mistake could arise with the organs so obvious as they were in this case." Croom chalked up the persistence of the mistake to the fact that the two siblings "as children were brought up abroad, in [the] charge of coloured nurses" who either did not realize, or did not mention, that something was amiss.[7] A relatively simple error had been committed and allowed to continue until the messy (presumed necessary) resolution was enacted:

> Many formalities had to be gone through; questions with regard to their father's will had to be adjusted, and new names had to be given to the young men, who donned men's attire, left this country, and are now pursuing useful avocations in the far East.[8]

A rather unusual brand of British export, to be sure. As this case concisely shows, sex diagnosis, no matter how simple, has never been a merely intellectual exercise.

Compared with their Continental analogues, British subjects of mistaken sex seem typically to have gone longer without suspicion of error. This was apparently due largely to what the editors of the *British Medical Journal* in 1885 lambasted as "the complete ignorance regarding the sexual organs and the sexual functions which is permitted, and, indeed, sedulously fostered, by the ordinary education which boys and girls receive in this country."[9] Although this editorial on "Sexual Ignorance" and subsequent responses did not specifically enumerate mistaken sex as one of the "serious evils" that resulted from the situation, they might have if their authors had consulted Lawson Tait's 1879 treatise *Diseases of Women*. There Tait specifically listed feminine ignorance in sexual matters as one reason why doubtful children should be raised male:

> By the time a male arrives at the age of marriage, he will have learnt, from the education which all men go through soon after puberty, whether or not he has marital capacity; and if he finds that he has not, he will not attempt to enter married life. But the majority of women enter the married state with but a very hazy notion of what

its functions are, a misfortune to which a large proportion of their special diseases may be attributed. If a malformed male, therefore, should be brought up as a woman, he may enter, and in very many instances actually has entered, the state of marriage, utterly unaware of his misfortune.[10]

By contrast, rather than blaming sexual secrecy for the spread of mistaken sex, French hermaphroditism specialists typically denounced "ignorant matrons," "prudish midwives," and "myopic doctors."[11] Hermaphroditism experts of both nations hoped that through their work they could at least educate medical practitioners about the problem and so reduce the number of cases of mistaken sex, although the experts preferred to be consulted personally.

Throughout the nineteenth century, medical men and doubtful-sex subjects hoped and assumed consultation with one or more experienced medical men would help resolve their questions. On numerous occasions medical men around Europe met in groups—presuming as they did that there was strength in numbers—to try to alleviate doubt in individual cases of doubtful sex. The stories of Marie-Madeleine Lefort and Marie Rosine/Gottlieb Göttlich were early instances of such attempts to determine what we might call consensus-based sex—that is, attempts to come to a professional consensus about a doubtful person's "true" sex. All over Europe, Britain, and the United States, medical practitioners also gathered in more informal groups to seek one another's advice in cases of doubtful sex. In England just before the turn of the century, following an initial examination of a "woman" patient, the Sheffield surgeon G. R. Green was still "puzzled" about the case, so he suggested an examination under ether, and the patient agreed. She was admitted to the Cottage Hospital and examined by Green and five summoned colleagues. Apparently in this case they all agreed she was male.[12] Practitioners faced with confusing cases also consulted their expert colleagues in lieu of large-group deliberations. In 1879, for example, Lawson Tait examined two children sent from Turkey specifically for the purpose of having their sexes diagnosed.[13]

As the century progressed and more medical societies formed, the frequency and formality of doubtful-sex gatherings seems to have increased. In November of 1875, in one such instance, Dr. De Gorrequer Griffith brought a case of doubtful sex in a living person before the Medical Society of London and sought its members' advice. Those present could not agree on the likely sex, and they appointed a committee to pursue the matter.[14] Sometimes a consensus did emerge, as when, in 1882, a "child was considered a female by the meeting" of the South Eastern Branch, East Kent District, of the British Medical Association.[15] In Paris in 1886 Alcide Benoist brought the case of Joseph-Marie X., formerly known as Marie X., before the Société de Médecine Légale de France. Benoist was part of a three-member medical panel officially appointed to confirm that Joseph-Marie should be legally declared a male. They did confirm his masculinity, but notably, at least one colleague at the meeting was distressed by the conclusion of the panel, and that member confessed he thought that "the tribunal ought to be very troubled in the face of such conclusions which seem to me imprecise."[16] In Britain as in France, presenting one's case and diagnosis often meant facing strong challenges to that diagnosis; recall the debate Barnes's case stirred up.

In fact, as we will see again and again, consultations with fellow medical men almost invariably, rather than clearing up confusion, resulted instead in deeper and broader doubt. Rather than finding strength in numbers, medical men often discovered that too many diagnosers spoiled the certainty. Remarkably little agreement emerged on the true nature of sex.

Still, as we will also see, even though there was astonishingly little agreement about which traits were significant or necessary to malehood or femalehood—about what "sex" was—there seems to have been almost universal agreement that femalehood and malehood were two very distinct forms of sex, and that they were the only "natural" forms of sex for humans. Practitioners seem to have invariably assumed that their uncertainty applied only to single cases of doubtful sex—even only to single organs. Never do they seem to have considered what we will see from a review of dozens of these discussions of doubtful sex: that the doubt extended far beyond any individual case to the endeavor as a whole.

Neugebauer and nearly all his colleagues invariably initially assumed that patients of doubtful sex were victims of "spurious hermaphroditism." That is, ambiguous subjects were presumed to be individuals in whom a single "true [male or female] sex is masked or rendered dubious by the existence of malformations."[17] Summarizing this hopeful, general medical opinion of the time, the American Theophilus Parvin remarked in his 1886 textbook *The Science and Art of Obstetrics,* "In almost all cases of alleged hermaphrodism this condition is *apparent,* not *real.*"[18] This diagnostic supposition seems to have been due partly to collective medical experience—under any definition of "true hermaphroditism," very few people were ever classified as such—and partly to pragmatism—no one knew quite what to do with a living patient who was not a man or a woman exclusively. Medical men presumed therefore that every doubtful subject could, should, and would be shown to be "really" a woman or man in whom confusing traits masked a single "true sex." The mission of the medical practitioner faced with doubtful sex was thus not to tackle the huge problem of the nature of sex—something that seemed in fact not to be a problem at all—but simply to get down to the case at hand and to sort the merely "apparent" from the "true": to seek out and recognize the veritable vulva.

Finding and Weighing Signs of Sex

Lawson Tait declared in 1879 that, in cases of hermaphroditism, "the glandular element must always be considered as the chief element of sex," and in fact it routinely was in the later nineteenth century.[19] In Britain as in France, in a doubtful situation medical men unanimously agreed that if one could find conclusive evidence of ovaries or testicles in a patient, the question of true sex was solved. Thus when Frederick Willcocks, M.D., examined a child named Rebecca brought to the Evelina Hospital for Sick Children in April of 1884, he determined that her gonads "were obviously testes, and the diagnosis was accordingly made that the child was a mal-formed male."[20] Conversely St. Thomas's Hospital Museum's specimen number LL-99, from a premature fetus, was "reckoned female in sex, since the generative glands consist[ed] of two well-developed ovaries."[21] As one participant in the 1888 British Gynaecological Society discussion remarked, whenever diagnosing doubtful sex, investigators preferred to be able "to say definitely . . . they had found ovaries or testes."[22] Practitioners burdened with doubtful children often waited to render definitive opinions, as they "look[ed] forward to the period of puberty" with the hope "that all doubt [would] then be set at rest."[23] Specifically they hoped for some new and definitive sign of the nature of the gonads, such as the characteristic descent of testicles.

The gonads, then, stood as "signs destined to level doubts in doubtful cases."[24] And no one disagreed in principle with this at the 1886 discussion on doubtful sex at the Société de Médecine Légale de France occasioned by Benoist's presentation of the case of Joseph-Marie X. In fact one participant argued specifically that "in order to characterize the sex, it is necessary to find testicles or ovaries. If one cannot find these organs, it is impossible to pronounce the sex of an individual." But another member of the society, one who had by then seen several such living doubtful cases come before the group, countered that such certainty was a luxury, that often, "in cases of teratology, testicles or ovaries are difficult to find."[25]

In practice the failure to find particular organs, such as ovaries or testicles, was sometimes taken as evidence that the organ did not exist, as when Barnes concluded that his patient did not have ovaries because he could not locate them. Nonetheless, many were wary of such reasoning by absentia, as was C. E. Underhill when he admonished the Obstetrical Society of Edinburgh in 1876, "In the living woman absence of uterus cannot with certainty be recognised." Underhill was specifically concerned that practitioners "avoid confounding these cases [females with no apparent uteri] with male hermaphrodites" who have "the external organs and formation of the female along with misplaced testicles."[26] Even if a practitioner did find an organ, how could he be sure it was what it seemed? In actual examinations, figuring whether an organ was an ovary or a testis—especially in a living, unopened patient—could be extraordinarily difficult. Testicles might remain undescended, and ovaries might be founded in unexpected places, too.[27] Neugebauer warned that "an ectopic ovary has often been taken for a testicle, or a testicle delayed in its descent for an ovary."[28]

Indeed, combinations of "deceptive" organs made matters utterly confusing. Neugebauer noted that underdevelopment of the penis combined with undescended testicles

and the development of a more or less rudimentary vagina simulates the presence of a vulva; hypertrophy of the clitoris simulates hypospadias of a penis; labial ectopia [displacement] of the ovaries, the testicles; adherence of the labia majora, the scrotum; non-adherent labia, a divided scrotum.[29]

Navigation of this sea of doubt was no job for the hasty. The sexes could look and feel remarkably similar even to the experienced medical man.

With all these confusing possibilities, no wonder that in 1893 John Lindsay, a Glasgow medical man, cautioned that "to found an opinion on the external appearance alone would be in the highest degree unscientific."[30] Lindsay personally faced an especially difficult situation, a triple dose of doubtful sex in three young siblings. For several years the parents had successfully hidden the ambiguity of Robert, Jessie, and Lizzie—aged seven, six, and four—until Lindsay, called in to attend the family during a bout of measles, discovered the situation. Lindsay thought he could discern that the children were all male, and yet he knew only too well that "the instances in which the *post-mortem* dissection of such cases has falsified the *ante-mortem* diagnosis are numerous enough and striking enough to make one doubt whether during life an opinion as to the sex can be given with any assurance at all."[31] For Lindsay, exploratory surgery was out of the question given its risks, and besides, it would not necessarily be helpful; biopsies were not performed on hermaphrodites' sex glands until the 1910s, and *in vivo* macroscopic visual inspection might not reveal the nature of the tissue of the gland and therefore, by implication, the sex.

Yet both Fancourt Barnes, who readily admitted "the difficulty of deciding these cases without a post-mortem examination,"[32] and John Lindsay, who confessed that "the whole story of sexual malformations make[s] one venture on a diagnosis with much diffidence,"[33] felt it necessary to provide "strong opinions" of their patients' sex, even at the risk of being "unscientific." Barnes, Lindsay, Pozzi, Neugebauer, and virtually all the other medical men so confronted with ambiguous sex reasoned that if, stifled by their doubts, they let a potentially mistaken sex go without at least expressing their well-informed opinions, they might allow an "error of sex" to continue, ultimately engendering wrongful occupations, scandalous unions, and broken lives.

How then did they propose to determine the sex? By doing what the British Gynaecological Society tried to do for Barnes's "living specimen": by "sum[ming] up the whole of the evidence," by evaluating and calculating the symptoms of sex until the "preponderance of evidence" weighed in favor of one.[34] Certainly, Lindsay conceded, given such young subjects as his own, one might think he could

> look forward to the period of puberty with the assurance that all doubt will then be set at rest. The descent of the testicles at that time would be almost conclusive. But there would be the difficulty of being sure that the descending bodies were really what they seemed to be, a difficulty that might be great enough were these, as they often are, atrophied or otherwise abnormal. Then the testicles might not descend at all, and in any case regard to the general habit of the body

would probably add to rather than diminish the confusion. These individuals, in respect to the secondary sexual characters, are frequently of an indifferent cast, or show a mingling of characters proper to the sexes separately. At puberty, as in infancy, the classification of those who are sexually abnormal turns on the estimation of probabilities.[35]

Lindsay meant this partly in a literal sense, that a practitioner ought to know that the record seemed to indicate that an ambiguous patient was much more likely to be truly male (have testicles) than truly female (have ovaries).[36] But Lindsay's "estimation of probabilities" also referred to the general belief held by physicians and surgeons that sex could often be estimated by calculating the many possible signs of true sex.

And these were many, for even while medical men insisted that the ovaries and testicles were the single and final markers of true sex, they still retained the assumption that those gonads would somehow make themselves known to the investigator through other traits, if only the investigator could successfully sort the true from the spurious signs. It seemed to many, for instance, that the voice of the subject would literally bespeak the true sex.[37] (Recall that Barnes's "own reasons for believing" his "living specimen" to be a male included "the *timbre* of the voice.")[38] A voice which sounded "hoarse" and "rude" was figured to be that of a man.[39] Indeed, a loud, harsh, or deep-toned voice seemed strangely incongruous in a woman.[40] The "true voice of woman" rang "sweet and agreeable,"[41] indeed "almost tender"[42] by nature, rather like that of a child.[43] So particular to the two sexes were these two sorts of voices believed to be that one French practitioner fancied a masked sex calling out to him: "while turning my head, I listened to her speak, [and] I had the utmost difficulty in figuring to myself that I was not listening to a boy."[44]

So while testicles and ovaries might have been the essential signs of sex, it seemed they were never thought of as the only signs available to the practitioner. This was a main point in a lecture on the medico-legal problems of hermaphrodites given in 1887 by Paul Camille Hippolyte Brouardel at the Faculté de Médecine in Paris. Medical students had to be prepared to cope with doubtful sex, British and French specialists agreed, for "it might come before any practitioner at any time, to have to adjudicate as to the sex."[45] In preparing his students for hermaphrodites to come, Brouardel thus instructed that "if you research carefully all the characteristics . . . you [will] reunite nearly always such an ensemble of probabilities that it is the equivalent to a proof" of sex.[46] It was not enough to consider just a few of the auxiliary signs, however. For

> in a decision of this nature, lacking evident organs, one takes into account [all] the general signs offered by the subject, like the hair, beard, breasts, the development of the hips, the voice, the instincts, etc. Any of these taken alone will not entail the conviction, but considered in their ensemble, they are not devoid of value.[47]

Medical men therefore presumed—or hoped—that, even in the face of persistent doubt and disagreement, given enough clues, true sex was sure to be discernible. True sex would surely out.

Bravery, Modesty, and Other Telltale Signs

By no means did medical men limit themselves to anatomy and physiology when seeking signs of true sex. Many British and French medical practitioners shared an unarticulated assumption—presumably one that they felt was so obvious it did not need articulation—that in the behavior of any person, doubtful or undoubted, the true sex would naturally reveal itself. For instance, carnal desires and acts figured prominently as signifiers of true sex. The assumption that true males—even those with mistaken identity—would naturally desire females, and true females males, pervaded the medical literature on doubtful sex. (Issues of carnal desire are explored in the next chapter.) And, although an error of sex could oblige a hermaphrodite "to live in the company of individuals of a different sex and to wear clothes which are not agreeable to [his/her] aptitudes and tastes," still the aptitudes and tastes of the "true sex" were expected to rise to the surface and bubble through.[48] A practitioner ought not be fooled by womanly clothes when the hermaphrodite who came to him "had the whole allure, the unself-consciousness of a man in his gaze, his gestures, and his walk."[49]

We should note that today many people draw distinctions between "sex" (used to refer specifically to particularly male or particularly female anatomy and physiology), "gender" (thought of as a category of self or social identification), and "sexuality" (used to refer to sexual desires or acts). This distinction is a uniquely twentieth-century Western phenomenon in which "sex" is generally thought of as being correlated with "nature," that is, the "congenital" state of the male and female, and "gender" is generally thought of as being correlated with "culture," that is, the "social" shaping of the individual as boy or girl, man or woman. So, for instance (to use stereotypical examples), ovaries and testes are now thought of as features of sex, whereas aggressiveness and passivity are often thought of as features of gender. But at least until the very late nineteenth century, such a sharp, tripartite division did not really exist in the West.[50] The characteristics we think of as belonging to the categories of sex, gender, and sexuality were generally supposed to belong naturally together—even if some people seemed to violate the rules. In the case of Barnes's "living specimen," the medical men at the British Gynaecological Society meeting took into consideration traits that we now think of as "sex" traits in their search for the patient's "true sex," that is, they considered things like gonads and genitalia and breast size, but many of them also took into consideration signs that we now think of as "gendered traits" and "sexuality": Barnes, for example, offered information from his patient's mother that the questionable subject's "tastes were decidedly feminine, but she had never shewn any partiality towards the male sex. She was affectionate and gentle towards her brothers and sisters."[51] These features, which we might now categorize as matters of gender and sexuality, were clearly seen by many medical men as correlated with—as part of—sex.

Thus many practitioners held the belief that if a person lacked the "aptitudes" of his or her sex, there was reason to suspect that things might not be what they seemed—there was reason to suspect hermaphroditism.[52] Doubtful, for instance, was a little girl who "never cared for girls' toys, play, or society" and had the "dispositions and tastes" of a boy.[53] Suspicions were also cast upon those children who were ambiguous in their tastes as well as in their genitalia. During her childhood, a French girl named Adele "appeared to have simultaneously had the tastes of girls and boys. She preferred games of force and skill, like running in the woods, climbing trees. On the other hand, she adored dolls, especially one with the name of Zélie, for whom she retains the most lively memories."[54] Some suggested that if the true sex were acknowledged, all spurious tastes and habits would cease.[55]

Modesty and bravery were, respectively, sexed as female and male in both France and Britain. It was the true woman who, timid and shy, reluctantly faced the medical man.[56] A fifteen-year-old "boy" with swollen breasts "was very unwilling to be exposed, and behaved just like a girl" with "an expression of modesty which [the observer] never [before] saw in a boy or man."[57] By contrast, a French practitioner was "struck by the facility with which [a 'woman' named] Josephine lent herself to observation. I and numerous colleagues who examined her never noticed that sentiment of fear and of caution which exists in every woman."[58]

A move from life as a supposed woman to life as a true man would also see a change from modesty to confidence. With an almost biblical lyricism, Neugebauer told the story of one German hermaphrodite:

> The young person [in question] was merely a male in whom hypospadias was complicated by an inguinal hernia; and when informed by Berthold of his discovery, blushed to her forehead and in outraged modesty would not believe the Professor's statement. . . . Berthold had forgotten the case when seven years afterwards he received a letter begging him to assist his former patient in rectifying his social position, in order that he might marry a young girl who was his mistress. And the patient, formerly timid and shamefaced as a girl, at the latter period betrayed all the *aplomb* and assurance of a man, and appeared naked and not ashamed before a company of medical men.[59]

Still, not every case of doubtful sex bolstered the assumption that tastes and behaviors matched the true sex. Some children displayed signs of their apparent sex, and not their medical sex, or so their mothers claimed.[60] Of course the mothers could be wrong. That was assumed to be common enough. But sometimes observations by doctors confirmed a man, gonadally male, had "all the habits and tastes of a woman, the toilette, the conversation."[61]

A few medical men were therefore skeptical about using behaviors and aptitudes as signifiers of sex. In instructing his medical students at Paris, for instance, Brouardel disdained tastes and talents as reliable indicators of sex. With his usual ironic wit, Brouardel had this warning for his class:

> I have noticed that every individual who wishes to be declared a man tells me . . . that he has a very pronounced taste for all the exercises of the body, gymnastics, arms, horse-riding, that he has a very broad intelligence and above all special aptitude for math—I ask myself if

this is peculiar to hermaphrodites!—In a word, the statements are so exaggerated that I dare not take them completely into account, and I encourage you to not let yourself be dragged by the moral orientation of an individual, often deceptive.[62]

But not all practitioners were so strong as Brouardel not to "be dragged by the moral orientation of an individual" into a particular diagnosis of sex. In spite of much evidence to the contrary, the general assumption stood that these traits would match the true sex, naturally and invariably.

Tools of the Trade

Although testes and especially ovaries often did not present themselves physically in any obvious way, several techniques were available to the medical investigator who sought concealed organs. British and French medical practitioners often knew of and were willing to employ a wide spectrum of techniques in their investigations of doubtful bodies. These approaches spanned from the most internal, such as exploratory surgery, to the most external, such as examination of relatives to look for signs of deformity common to the family. The British seem to have been particularly inclined to trace and examine family in the search for similar deformities. One medical investigator reported, for instance, that he "went into [the family history] as thoroughly as possible, personally examining where feasible each individual."[63] Another tracked down all the living siblings of a hermaphroditic patient, and asked various colleagues to examine and watch them; two of the suspicious brothers were followed "so that in the future we may hear more of them, especially should menstruation occur through the penis."[64]

Obviously it was easier to thoroughly examine the patient at hand than to try to examine all of his or her relatives, too. Certain invasive manual examinations were therefore employed as a way of literally feeling out any telltale organs (ovaries, testes, uterus, prostate) that might lie hidden in the abdominal cavity. The success of these various examinations seemed to depend on the knowledge and expertise of the practitioner and the veracity and cooperation of the patient. Cooperation was a serious issue in many cases, especially when patients felt shame about their genitalia or deformities or discomfort with invasive examinations. Sometimes examiners administered a general anesthetic before these procedures in order to lessen the pain caused the patient's body or modesty.[65]

When a vagina or apparent vagina existed, examinations were performed on it to determine its length, width, texture, and whether or not it ended blindly or at a uterus.[66] Likewise bladder sounding and catheterization gave information on the length and direction of the urethra, and on the position of the urethra and bladder.[67] By about 1890 the rectal examination had become standard practice in the diagnosis of sex, although styles differed.[68] For his part, Fancourt Barnes thought that digital rectal examinations sufficed, and "he did not consider that it was justifiable to expose a patient to the risks and inconveniences inseparable from the introduction of the whole hand into the rectum."[69] Practitioners of rectal exams searched for evidence of ovaries, retained testes, uteri, vaginas, and prostates. Yet some question remained on the usefulness of such an exam.[70] Additionally several medical men reported undoubted "women in whom no uterus could be felt" using techniques that were supposed to reveal a uterus in any woman, so that the value of these sorts of examinations sometimes fell into question.[71]

By the early 1890s detailed descriptions of these physical examination techniques had reached French and British gynecological textbooks, which undoubtedly led to their more widespread use. Especially popular were simultaneous combinations of these examinations, because such combinations could help the practitioner hone in on internal organs. So, for instance, in the discussion of Barnes's case, Dr. Aveling suggested that "if the patient were put under an anaesthetic and the bimanual examination made, per rectum and through the abdomen, they might succeed in finding ovaries or testes."[72] In their *Traité Pratique de Gynécologie* of 1894, Stephane Bonnet and Paul Petit instructed:

> In order to diagnose profound malformations, one would use most particularly . . . the rectal examination combined with abdominal palpation, or a recto-vesicle examination [that is, a rectal examination combined with bladder sounding]. If needed, one has recourse to chloroform anaesthesia.[73]

All possible attempts were to be made to literally feel out the nature of the sex.

In his 1899 publication on the types of hermaphroditism, Xavier Delore, the medical doctor from Lyons, advised his colleagues that "if the superficial genital area does not give a satisfactory diagnosis on the nature of the sex, there is a new resource, namely *laparotomy,* which furnishes now an interesting and unprecedented element of investigation."[74] Yet the use of exploratory surgery (laparotomy) for diagnosing sex must have been recognized in France as early as 1892, because at least one practitioner found it necessary that year to defend his failure to perform exploratory surgery on a doubtful patient.[75] Regular use of this technique exclusively for the purpose of sex diagnosis did not become common, however, until at least the second decade of the twentieth century in Britain and France, when improvements in anesthesia and antiseptic techniques made it relatively safe and submission by patients made it easy. In any case, laparotomy could not always be relied upon, since sex glands were sometimes shown to look like one type and be another.[76]

Exploratory surgery was, of course, a relatively dangerous form of examination, and during the late nineteenth-century surgeons performed such procedures usually because of a need to investigate more pressing problems, such as acute abdominal pain. In 1895, for example, a surgeon in Paris performed exploratory surgery in an attempt to locate the cause of recurrent abdominal pain in a young doubtful-sex subject generally considered to be male. The Parisian surgeon discovered ovaries in the adolescent, and removed them with their fallopian tubes in order to "shelter this young woman from her periodic neuralgics."[77] (She had no vagina through which the menstrual blood might drain.) Similarly a surgeon in Glasgow could not discern what the bodies in the labia of Christina, his nine-year-old child subject, were, but "in any case he decided to cut down on the tumours of the left side," because if

> it proved to be an ovary it was in a very unsuitable position, for at the period of menstruation it would give rise to extreme discomforts, lying as it did in the labium, so that unless it could be slipped into the abdomen through the large inguinal ring he would cut it off. If it proved to be a testicle he certainly would excise it, as it would be a singularly unfortunate thing for a person with all the outward appearance of a female to develop into a man as to internal organs and feelings.[78]

It was a testicle, and he excised it and its mate. (The implications of this are discussed in the next chapter.)

Occasionally therapeutic surgery revealed the true sex of a subject even though the surgeon had conceived no doubts on that sex until *after* the incision revealed unexpected gonads.[79] Samuel Pozzi reported on just such a case of mistaken sex in which the mistake was discovered during an operation for a hernia: "If the appearance of the internal organs in a painful double hernia had not necessitated an operation, one would never have been able to suppose that this woman carried testicles, and consequently was a man."[80] One now wonders if medical men ever considered how many "true men" and "true women" might go forever undetected.

Tricky Genital Geography

Even without the aid of surgery, techniques were available to the medical investigator by which he might manage to elicit symptoms characteristic of testicles or ovaries from apparent gonads. So, for instance, when a forty-two-year-old widow was admitted into Bird Ward of Middlesex Hospital "on account of a painful swelling in the left groin," the attending surgeon, Andrew Clark, suspected the swelling was really a recently descended testicle because he managed to rouse classically testicular behavior from the swelling: "There was no impulse on coughing, it was dull on percussion, very tender to the touch, and not reducible into the abdominal cavity."[81] Physical examinations of ambiguous persons typically included the palpation and sometimes percussion of any lump or swelling in the scrotum or labia majora.[82] If "methodic pressure" were applied, sickening or painful feelings of various sorts might indicate testicles or hernias. So, for instance, methodic pressure applied in the case of "A." showed nothing like testicles, and "moreover A. never felt any pain in these regions, whether following violent exertions to which she was often given, or following prolonged erections which she has sometimes had."[83] By contrast another surgeon's patient suffered "sickening pain on pressure and handling"—signs of testicles.[84]

In an attempt to distinguish between a testicle and a nontesticular hernia in a supposed woman—lumps in apparent labia might be either—practitioners sometimes tried to elicit a cremaster reflex, that is, a spontaneous and sudden elevation of the testicle induced by certain stimuli like cold or an unanticipated stroke on the inner thigh. But this was yet another one of the many contested signs of sex. Many practitioners thought that elicitation of a cremaster-like reflex from a genital swelling definitely meant it was a testicle.[85] Not everyone, however, "believed the cremasteric reflex to be an absolute proof of the existence of a testicle,"[86] or the want of it to be proof of no testicle.[87]

Pressure applied to the genital region sometimes had the added diagnostic benefit of producing erection of the penis or clitoris, erection that could to the investigator signify either penis-like or clitoris-like behavior.[88] French authors seem to have been less easily persuaded that the clitoris and penis would act significantly differently, although they were generally convinced that uniquely feminine stimulation occurred mostly through rubbing of the vagina.[89] This vaginal rubbing was thought to ultimately result in the telling "voluptuous sensations and . . . secretion" from "the glands like those which one finds in woman."[90]

Still, many hermaphroditism experts admonished medical men not to trust the external genitalia or the vagina to bespeak the true sex. The external genitalia were often confusing or downright misleading. They were after all, in most cases of mistaken sex, the very cause of the error. What looked like labia majora might "really" be a bifid scrotum.[91] Some medical men thought the presence or absence of pubic hair on the inner surfaces of supposed labia majora settled the question of their status; presence of the hair on their inner surfaces would indicate a bifid scrotum.[92] (Indeed such hair alone was enough for one English doctor to refer to such labia as "pseudo labia.")[93] But not all agreed that the layout of pubic hair could be trusted any more than the rest of the genital geography. Hermaphrodites were especially infamous for possessing supposed clitorises that looked and acted like penises, and vice versa.

Even so, reporting practitioners typically provided colleagues with minute detail of a suspicious phallus's structure, length, width, fixture, density, erectile capabilities, and perforations.[94] A "well-formed prepuce and glans penis" marked with a functional *meatus urinarus* (urethral opening) typically signified a true penis—that is, an organ belonging to a true male.[95] But none of these characteristics could be relied upon as definitive. Penises might be imperforate or misshapen, as in some cases of hypospadias,[96] and a clitoris might be grooved or even perforated like a penis.[97] Size could not be trustworthy, either.[98] Although clitorises were expected to be significantly smaller than penises, practitioners found many cases of enlarged clitorises and atrophied penises.[99] Doubt remained as to the essential characteristics of the penis and clitoris, and while this doubt remained, any phallus made an unreliable sign. It was not uncommon for a medical writer to withhold judgment on such a confusing organ; he might simply refer to the troublesome item as "a clitoris or penis" or "the penis-like organ."[100]

Emissions of Sex

As we have seen, in actual cases of hermaphroditism, the diagnosis of sex often seemed to differ depending on the eye, the hand, and the mind of the professional beholder. In his efforts to make more rigorous and uniform such diagnosis in doubtful cases, Neugebauer advised his colleagues around Europe and Britain to go beyond mere manual prodding and poking and to search further for clues literally emitted by the testicles, because, he concluded from his long studies, "error of sex in a male hypospadiac [a true male spurious hermaphrodite] brought up as a girl reveals itself comparatively frequently by seminal ejaculations."[101] Nocturnal emissions and ejaculation during intercourse spoke to the subject's malehood. Thus just as true women were supposed to regularly menstruate, true men were supposed to occasionally ejaculate—although, if the relative levels of discussion in the literature were to be taken as any indication of actual activity, apparently true men masturbated and ejaculated much more often in France than in Britain.

French reporting by patients and practitioners of sexual stimulation and actions far outran British reporting.[102] Such discharges of sex were important enough to French medical practitioners that they often made a point of carefully eliciting detailed descriptions from patients. Indeed, French medical men tried to personally witness sexed emissions whenever possible. For instance, doubtful patients suspected of being men but who claimed to menstruate were often kept under observation,[103] and one French investigator found it quite opportunistic that a patient questioned about her sexual practices wound up stimulated as a result. That patient thus provided her investigator with a first-hand look at her erection and ejaculation. He recounted the serendipitous scene:

> In the course of my examination—psychological effect of the questioning arising from this order of ideas?—I found myself attended by the erection of the clitoris, which took on then the volume of a little finger, and by the emission of a secretion with the identical aspect of the prostatic liquid of a man, fluid, transparent, with a

slightly strong odor, and of which I was able to retain a small amount on a microscope slide

by wiping it with a piece of cotton.[104] He examined it, and it contained no trace of spermatozoa—spermatozoa were considered a very significant sign of malehood—but to the practitioner who personally witnessed them, this erection and ejaculation could act as persuasive signs of true masculinity in spite of the absence of spermatozoa. In 1892 a surgeon of Lille took as diagnostically significant that in his doubtful patient "there has never been the least appearance of the menses, whereas on the other hand there have been erections and ejaculations."[105] French authors like these noted with a tone of significance whether a patient ejaculated, especially if the fluid discharged "starched the line" on the clothes or sheets.[106] Again, even if the supposed semen were devoid of spermatozoa, such starchy, "characteristic semen" indicated a likely man.[107] So, as with all other signs, these—ejaculation, sperm, erection, menstruation—were not simply positive or negative signs of sex, but indicators of *possible* sex, indicators that had to be screened and weighed by each interested party against his own criteria, to judge reality and significance.

The Pain of Menstruation

In matters of hermaphroditism practitioners often relied on the menses as the female diagnostic counterpart to semen. "The presence of menses seems to be a very evident proof in favor of the feminine sex," or, "if not a proof, at least a great probability in favor of the feminine sex"; so commented one French medical observer of sex in 1895.[108] Indeed menstruation seemed so very real a feminine trait, that the very absence of it was something to be "seen"; the absence of it seemed to have a presence as a sign.[109] But like so many of the traits generally thought sexed, menstruation as a sign was rife with problems, problems that boiled down to this: some women did not menstruate, and some men did.

For his part, Neugebauer was greatly distressed that his colleagues did not more often consider mistaken sex as the explanation for persistent amenorrhoea (failure to menstruate) in supposed women. He reminded colleagues that "the practitioner, whenever he is confronted with a case of absolute amenorrhoea, should always remember the possibility of such a mistake [of sex]. Under such circumstances to omit an examination of the genital organs out of consideration for the timid modesty of a young girl, may be a matter of great regret."[110] Neugebauer insisted that amenorrhoea was the condition that "more frequently than any other . . . led to the discovery of an error of sex."[111] But as even Neugebauer knew, amenorrhoea was not by any means a sure sign of mistaken sex. Failure to menstruate might signal an underlying male, a mistaken sex,[112] however, amenorrhea could also signify a genuine woman troubled by, for example, imperforation of the vagina or congenital absence of the uterus—or something as common as pregnancy. Although some practitioners did indeed use absence of menstruation as a sign of sex, absence of menstruation often signaled a diagnostic problem, not a solution.[113] It did not necessarily reveal an absence of ovaries.[114]

Conversely, it was not clear that the presence of menstruation positively indicated ovaries. Some medical practitioners testified that they had seen women without ovaries menstruate. The French gynecologist Magitot thought "with regard to the menses, that they reappear often after the ablation [destruction] of the ovaries, by virtue of a sort of physiological habit, a sort of acquired force."[115] Clémence Royer debated this very point with Samuel Pozzi at a meeting of the Anthropological Society of Paris in December 1889, on the occasion of the presentation of a case of hermaphroditism. Royer—notably the only woman we come across on the professional side of these doubtful-sex discussions—doubted the "necessary relation which is said to exist between the presence of ovaries and the regularity of menstrual flux." She knew personally of cases in which the ovaries had been surgically destroyed, and "the menstrual flux [had] reappeared just one month later, on the accustomed date and perfectly regular, only with a little less abundance." But Pozzi responded that this was not the rule in such cases: "There is generally irregularity at first, then disappearance at the end of a certain time."[116] Both Royer and Pozzi supported their claims with case histories and authorities, and neither party would relent his or her opinion. In Britain, similar disagreement existed on this question of the necessity of ovaries to menstruation.[117]

To complicate matters still more, it was not clear what symptoms should count as menstruation. Some but not all practitioners were willing to take as signs of menstruation, in the absence of vaginal bleeding, periodic pains, swelling, and "vicarious" bleeding through nonvaginal orifices like the nose. Some thought there existed a number of "periodic troubles which [would] qualify as *compensators* when the menses are not manifested by a bloody utero-vulvar discharge."[118] Consequently, for instance, although his subject had never had monthly bleeding, Sir Hector Clare Cameron, professor of clinical surgery at the University of Glasgow, grew suspicious of the sex of a young man burdened monthly by excruciating attacks of abdominal pain. They had started when the patient was thirteen years of age (he was now twenty-seven), came monthly, and lasted two or three days. Furthermore, "during each attack of pain the breasts became full and enlarged, and were so very tender that he could hardly tolerate the weight of the bedclothes upon them."[119] This certainly smacked of menstruation to Sir Hector, and sure enough, upon operating he discovered an ovary, the removal of which ended the complaints. Symptoms such as periodic pain of the abdomen or breasts, swelling of various body parts, and extravaginal bleeding were lumped under the rubric of "menstrual molimina," or auxiliary signs of the body's attempt to menstruate, that is, auxiliary symptoms of femininity. Like Sir Hector, many medical men did not require genital bleeding for menstruation, but not all accepted these "menstrual molimina" phenomena as definitive menstrual signs.

Moreover, just as some practitioners did not view periodic genital bleeding as a necessary sign of menstruation and therefore femalehood, several also did not consider it a sufficient sign of femalehood. The reason was simple enough: "true" men sometimes menstruated. Experts were at a complete loss to explain why it was that occasionally menstruation occurred in male spurious hermaphrodites, that is, womanish-looking men complete with testicles and no ovaries, like the Middlesex Hospital widow.[120] Certainly some of this could be chalked up to trickery, as in the infamous case of Catherine Hohmann in Germany (later married as Charles Hohmann in America), who was suspected of smearing herself with blood to feign menstruation. Hohmann and other hermaphrodites were sometimes closely monitored by medical men, doubtful of their menstrual veracity, and practitioners were even warned by a German colleague of "the great tendency these individuals have to deceive any medical man."[121] As noted in the last chapter, a few hermaphrodites, including Catherine Hohmann, profited by exhibiting themselves to medical men and other curious parties. The more profoundly mixed their sex appeared, the more they could earn, so Hohmann and others had a good reason to fake sex signs. Consequently, in some cases of apparent hermaphroditism, the patient's word was as doubted as the sex.[122]

Nonetheless, all trickery aside, some otherwise undoubted men appeared genuinely to menstruate.[123] "A.B." and his wife consulted William Rushton Parker because

> they had both noticed that he had periodical red monthly discharges, usually lasting about three days, and staining his shirts and sheets, they began to note the dates on which this began, and ticked off the following days: April 24th, May 22nd, June 19th, July 21st, August 14th, September 10th, October 9th, November 1st.[124]

The wife especially found the whole situation quite disconcerting, and she wanted to know of Rushton Parker "whether he [A.B.] was a man, or a woman, or both, or neither." The doctor guessed A.B. was "at least not a complete man," but was at a loss to give any more definite diagnosis. Like amenorrhea in women, menstruation in men was by no means considered a normal condition, but neither was it considered an inconceivable one. It was not at all clear a person needed ovaries and a uterus to menstruate.[125] Consequently Rushton Parker could not be sure what A.B.'s bleeding meant. The nature and meaning of menstruation was murky enough that many medical men would have agreed with Dr. Bantock's frustrated remark in the 1888 British Gynaecological Society discussion: "Menstruation again was hardly reliable."[126]

Of Woman Born: Breasts and the Pelvis

Ultimately the breasts, too, were found to be "hardly reliable" as indicators of true sex. The nineteenth-century Western mind often equated women with their reproductive functions, so it is not surprising that many medical men initially thought they could garner clues as to true sex from the sort of breasts and pelvis a person carried. The French waxed particularly poetic in their descriptions of the breasts of doubtful patients. Not uncommon was the sort of account provided by Auguste Lutaud in 1885: "[S]he [the subject] is provided with two very beautiful and very firm breasts in which one feels very distinctly the mammary gland and [the breast is] supplied with a pink and erectile nipple."[127] Most, if not all, French and British medical men during this period maintained similarly definite ideas about what made a breast feminine. "Feminine" breasts were those that were full, rounded, "with large nipples surrounded by a dark areola. They gave to one's hands grasping them the distinct feeling of mammary glands, and not merely that of masses of fat."[128] The presence of a critical mass of glandular tissue struck many as the most important aspect of the womanly breast,[129] but it was not enough that "the breasts were well developed" if "the nipples were rudimentary and the areolae were not marked."[130] Undeveloped breasts could throw suspicion on a woman as they did in the case of Sophie V., who had had to confess that her husband had breasts "much greater than she."[131]

The problem with using breasts for sex determination, however, was that men occasionally possessed these "feminine" breasts. In fact, the condition was documented often enough that it earned its own name, gynecomastia (or gynecomazia), defined in the late nineteenth century as the coincidence in a man of "feminine" breasts and shrunken, absent, or misplaced testicles. This condition constituted one of the many "masks" to true sex in male spurious hermaphrodites, but it was also known to afflict pubescent boys, obese men, male inebriates, and "idiots."[132] Sizable rounded breasts, however widely considered a feminine trait, still made an untrustworthy marker of sex, as they could indicate several pathologies other than true womanhood. Accordingly Samuel Pozzi insisted that "the *absence of breasts* signifies little; if they are, it is true, rarely lacking in women, they can nevertheless be missing exceptionally," and, besides, "one has seen men with feminine mammae."[133] Pozzi complained that, like some medical men, "the public wrongly attaches an excessive importance to this element ['feminine' breasts], entirely secondary in the determination of sex."[134]

As we have seen, few traits besides the gonads managed to remain entirely sex-distinctive, but the pelvis did ride the seas of sexual doubt relatively unscathed. Many charged with diagnosis of sex thought that surely the pelvis would not lie, there being such "striking differences in the size and shape of the pelvis in the two sexes."[135] A feminine pelvis was "ample and flared,"[136] "large" and "dug out,"[137] ready to receive and bear a child. A masculine pelvis was, by comparison, narrow and straight.[138] Believed as it was to be an ultra-sexed part, measurements of the pelvis seemed to make a handy diagnostic tool. In a dramatic demonstration of how firm could be this belief in profound pelvic sex difference, John Phillips, staff physician at the British Lying-In Hospital and the Chelsea Hospital for Women, measured the pelvic bones of an ambiguous baby in an attempt to determine its sex.[139] Phillips's methodology was unusual for diagnosing a baby, but not for resolving doubtful sex in adults. Medical reporters often employed pelvic measurement to reach and then provide evidence for their conclusions.[140] After about 1880 few writers bothered to provide explicit comparative charts of the pelvis sizes of doubted and undoubted subjects.[141] But they did often, in the body of their texts, provide minute detail on the size and shape of doubtful patients' pelvises, as when Lawson Tait described his encounter with a prisoner others thought male, whom he suspected female: "I obtained a photograph of her naked; and the outlines of the figure, having the wide pelvis, narrow chest, and inturned thighs, quite confirmed my opinion." She was, Tait became convinced, a female.[142]

Indeed, the entire skeleton seemed to stand as a literal framework that stayed sexually secure, no matter what the weight of the ambiguity hung upon it. In a lecture on the medico-legal aspects of doubtful sex at the University of Aberdeen, Francis Ogston, M.D., and professor of medical logic and medical jurisprudence, enumerated in vivid detail the structural contrasts between man and woman. He focused his lengthy sex-difference observations on the maxim: "the female is usually not so tall as the male, and the whole figure makes an approach to *the oval,* not witnessed in the case of the masculine form."[143] So the doubtful body's proportions allegedly would betray its sex. Men's limbs and appendages were assumed to be naturally more "stout and strong,"[144] and "large hands" conveyed a "masculine appearance."[145] The "rounded contour of the female form" was supposedly made possible by the woman's peculiar bone shape and fatty padding.[146] The curve of her hips pressed her knees together, while the prominent size and shape of his throat revealed what some thought that absolutely foundational organ of malehood—the conspicuous Adam's apple.[147]

In discussions of doubtful sex, women were said by medical writers of the late nineteenth century to be made generally of finer, lighter bones than men.[148] Consequently "large feet, enormous hands, strong development and [large] breadth of the shoulders" in a woman were suspect[149]—unless, of course, she was of working stock. And even if she were of working stock, a particularly strong woman was a glaring anomaly.[150] A true woman's shoulders sloped gently down from her neck,[151] and that neck was slender, without prominent thyroid cartilage or musculature.[152] The French seem to have been almost uniformly persuaded that adult height depended on sex. French reports of hermaphrodites are rife with height measurements of strangely tall "women" and suspiciously short "men."[153]

Just as the testes would reveal themselves through seminal emissions, allegedly the "feminine" pelvis would invariably betray itself by the production of a necessarily "feminine" walk: "the thighs near each other, the knees a little inside, with a slight waddle."[154] Even womanly clothing could not hide a characteristically masculine stride. In the case of one French hermaphrodite, "the general aspect, the movement and the walk easily put her in suspicion of [being] a sex other than the one indicated by her womanly vestments."[155]

It is important to note that, although phrases like "aspects" and "demeanors" generally seem to have referred to bone structure and body shape, it is sometimes impossible to ascertain what the medical man of the late nineteenth century saw when he declared that a "child looked more a girl than a boy"[156] or that a person had "a demeanor not characteristic of either sex, but inclining to that of a female."[157] The same is true for other sorts of traits. For instance, what might it have meant to say that a doubtful subject "had some predilection for the female sex, but no sexual desire"?[158] Such statements do indicate that there were assumed to be standard male and female types, but do not always make clear what those types were supposed to be. The implication was, however, that the educated (presumably male) reader of the medical report would know the sex standards.[159]

Hairier Problems Still

We can see that in discussions of breasts and pelvises, French and British medical writers generally took the masculine as either unproblematic or somehow characterless. For while comments like "the pelvis was masculine" abounded, rarely was that masculinity articulated. By contrast, medical men expounded upon the feminine breast and pelvis in deep and rich detail, as if the very nature of the pelvis and breasts was feminine, as if they constituted mere, unremarkable vestiges in the undoubted man.[160] (Or maybe they just enjoyed an opportunity to talk about women's breasts and hips?) By contrast, a more symmetrical level of description was recorded for the sexed traits of the face. A typical "feminine" face was said to possess "some regular, delicate traits, with a complete absence of a beard."[161] Conversely, "strongly accentuated" features and facial hair marked the masculine face.[162] Some practitioners relied partially on these facial signs to guess at sex.[163] Yet facial features constituted one of the most highly contested classes of signifiers. For while some medical men believed that no style of coiffure could "change . . . the really virile aspect of the traits of the face,"[164] others thought a feminine fashion of hair would necessarily and deceptively lend the face a certain false "feminine air."[165] To the latter camp, facial characteristics were "unsafe . . . as a guide, the dress making such a very great difference."[166]

What then of the issue of hair—that on the body, the face, or the top of the head? To many investigators, like voice at first hearing, hair at first glance seemed a good way to tell the boys from the girls, and especially the men from the women. Women supposedly had more hair up top, and less the rest of the way down.[167] Indeed some British writers seem to have assumed that head hair in a male would naturally stop growing before ever reaching the shoulders.[168] The French, however, were less apt to hold this conviction, faced as they often were with luxuriously long-haired male hermaphrodites. (One French medical practitioner in 1899 summarized a case of mistaken sex thus: "This large girl in long skirts and opulent hair was a boy.")[169] Still, medical men of both nations assumed masculine head hair to be qualitatively different from the feminine, the latter fine and silky by comparison.[170]

Similarly, men's trunks and limbs were supposed to be covered with coarse and profuse hair, so much so that this sort of hair on the thighs was deemed "masculine hairs."[171] Womanly skin was "white, fine, devoid of hairs,"[172] truly "delicate and fine."[173] Hair was of course allowable in the pubic region of the woman, but not if it reached to the umbilicus; such creeping pubic hair was thought a specifically masculine trait.[174] The pubic hair on a woman should be confined to the telltale, delicate triangle. Yet again, perhaps inevitably, the significance of the pubic hair was one of the many contested points, as it was, for example, at Barnes's 1888 presentation. At that meeting, while Dr. Aveling maintained "that hair did not grow there in the congenital absence of either testicles or ovaries," Lawson Tait disagreed. The latter cited a case of his own in which "there was no uterus and no ovaries, and the pelvis was like a tea cup, perfectly smooth and empty . . . yet the patient had hairs in abundance on the pubes."[175] Aveling and Tait never did come to a consensus on the significance of pubic hair or, more precisely, of its absence. This constituted just another example of the many in which an argument over a patient's sex shifted to the meaning of the signifiers themselves.

Facial hair in a supposed woman could be quite striking, as in a case reported by the Frenchman Accolas in 1889, in which an alleged woman's "profile recall[ed] a little that of Christ." Her beard was "silky, long, curly, reaching 0.12–0.13 meters. Under the jaw the color is dark brown; the moustache, less developed than the rest of the beard, is less dark in color and drawn lightly toward red."[176] Yet even though no one took special note of bearded men as they did bearded women, the latter were hardly unheard of. Facial hair presented itself as an especially problematic sign of sex. The meaning of facial hair—and its commonality in females of certain racial groups or classes—was a major point of contention in the discussion of Barnes's case, and its reliability as a signifier was equally contested by dozens of other medical authors. Indeed, one author launched into a veritable natural and social history of beards in his review article on feminine pseudohermaphrodites in a spirited attempt to demonstrate that true women were often bearded.[177] Still, some continued to credit it with great importance as a sex sign, enough so that Samuel Pozzi felt obliged to remind his colleagues often that "the *beard* as one knows can sometimes ornament the weaker sex."[178]

Mitigating Factors

To most practitioners a natural lack of facial hair signaled womanhood, or at least incomplete manhood.[179] Even if beards and moustaches were well known in women, they were still considered truly "manly" traits, and so required explanation when found on a woman. In Britain, facial hair in women was likely to be correlated with insanity.[180] For instance, one English practitioner remarked on the occasion of reporting a bearded woman, "It is not uncommon to find hirsute females amongst the insane; but all who have seen this case (including the Commissioners in Lunacy) assert that they have never before met with an insane woman having so perfect a beard and moustache as M.C—."[181] In France, abundant facial hair in women was more apt to be considered a sign of remarkable virility (here meaning strength), particularly notable among women of the countryside. A French provincial woman was married at seventeen, and her beard began to grow three years later, but her "husband was not very troubled by it; the beard was shaved Sundays for going to mass, and the spouses were vineyard workers [so] the muscular force of the bride was not to be disdained."[182] Insanity and strength were only two examples of several accepted mitigating factors. Race was also thought possible of altering the patterns of this otherwise "sexed" trait. In the 1888 British Gynaecological Society fracas, Routh's objection to Barnes's reliance on facial hair as a sex symptom indicated that Routh thought Jewishness enough to make facial hair grow on women. ("Many Jewesses had quite a large quantity of both beard and moustache.")[183]

This idea of allowable mitigating factors is important. Nineteenth-century ideas about sex difference look uniform on the surface because the doctrine of two sexes was so obvious and so often declared. But within that doctrine there was clearly much flexibility (and, as we have seen, astounding amounts of disagreement). Apparently many practitioners and lay people believed that the nature of several potentially diagnostic "sex" characteristics could be traced to nonsexual attributes. Thus, sex was technically dual but its multilayered nature was accepted by many. More examples: age and premature aging were believed by many to affect "sex" traits; young men might not have much or any facial hair, yet hairy jaws and upper lips were positively expected of postmenopausal women.[184] Meanwhile, infancy in males was sometimes believed to hide their telling organs, such that they appeared feminine, and old age was said often to incite in women "a manly appearance." One author related that he had

> had occasion to observe numerous examples of this in old women who have been long in prison, or in a lunatic asylum; in these the female breasts quite disappear, a beard is developed on the lips and chin, and they acquire a rough and manly voice, so that when lying in bed covered up to the chest, they may very readily be mistaken for a man.[185]

Conversely, pubescence was thought to often bring about feminine breasts in a male. Race, meanwhile, constituted a factor that could mitigate the nature of the breasts just as it mitigated facial hair; African women were thought to have significantly more pendulous breasts than European women.[186] And women of some Asian and African races were supposed commonly to possess hypertrophied clitorises and other "masculine" genitalia.[187] Onanism, too, was alleged capable of making any woman's clitoris hypertrophy to penis-like dimensions.[188] Race, age, sexual practices, insanity—all these and more could determine traits otherwise simply "sexed." Various kinds of education, climactic conditions, and occupation might also make a man womanly or a woman manly.[189]

Corraling Doubtful Sex

As we have seen, close analysis of the literature on hermaphroditism reveals that there was in France and Britain by no means a single, unified medical (or lay) opinion about which traits must be taken as essentially or significantly male or female. The sexes were constructed in many different—sometimes conflicting—ways in hermaphrodite theory and in practice. Consultation with colleagues did not necessarily resolve problems, as it would in a scientific ideal. Frequently it increased them, because it revealed latent disagreements about the nature of sex and its signs. For instance, the seemingly simple question, "Was there a uterus?" could explode into dozens of troublesome related questions: How was one to verify a uterus? If there was a uterus, would it necessarily be felt? What was the proper way to do a rectal exam? Should a colleague's interpretation of the exam or an analogous case be trusted? Should the patient's claim of menstruation be trusted? Should menstruation itself be trusted as a sign? Even if there was a uterus, could that be taken as conclusive in terms of sex diagnosis?

It was not uncommon for editors of French and British medical journals to employ the phrase "Doubtful Sex" to head a notice of a single case of sexual ambiguity. But this phrase could just as well have referred to the whole endeavor of sex in discussions of ambiguity, doubtful—literally full of doubt—as it was. Yet paradoxically, even those medical men who repeatedly faced hermaphrodites unfailingly maintained a fundamental belief in two separate and unequal sexes.

How then did medical practitioners, so unable to agree in practice on the nature of malehood and femalehood, manage to avoid being swallowed up in all the doubt and disagreement? We can tease out five basic strategies employed by practitioners faced with doubtful sex, strategies that achieved twin goals, namely, alleviation of doubt in any given individual case, and simultaneous forestalling of leakage of that doubt onto the broader issue of the basic natures of the sexes.

First, in any single instance of doubtful sex, medical practitioners searched the person in question for every possible sign of sex, within and without the doubtful body, in its anatomy, physiology, demeanor, tastes, and talents. The idea here was that the "true sex" of every body would ultimately be found and settled. Sex demands were pressed upon the hermaphrodite's body and thereby amplified, and any person successfully removed from the doubtful realm automatically added to the strength of sex distinctions.

Second, medical practitioners consulted with colleagues in the hopes that power against doubt could be found in numbers. Even though in practice these consultations often resulted in disagreements about sex (individual and global), the sheer energy devoted to the cause of controlling hermaphroditism made it look like it was being successfully handled. Moreover (and as we will see especially in Chapter 5), the very intellectual investigation into hermaphroditism—its domestication into the world of science—seemed to rein in doubtful sex from the realm of uncertainty, even if in fact huge doubt continued.

Third, practitioners found it necessary to doubt hermaphrodites in order to deflate the doubt the hermaphrodites seemed to throw on "natural" sex differences. Thus, for instance, the intellectual quandary caused by the menstruation of an apparent man could be solved by simply doubting the man's claim that he menstruated. The veracity or observational powers of the subject—and especially of subjects' mothers—were always open to question, and were often used as the way out of intellectual dilemmas caused by hermaphrodites.[190]

Fourth, in order to maintain two large categories of sex, mitigating circumstances which could change sex locally were allowed by most medical men. By explaining a sex-trait deviation, the idea of consistently "normal" sex traits was upheld. For instance, as we have seen, pubescence and obesity could be blamed for apparently feminine breasts in a male, race could mitigate the nature of the breasts and hair and other signs that were otherwise typically considered sex-determined, and so on. Explainable, otherwise seemingly "exceptional" cases were thus used as exceptions to prove the rule of two distinct sexes.

Fifth and finally, medical men labeled masculine traits in a true woman and feminine traits in a true man "spurious" or "false," robbing them of their meaning, imbuing them with the status of mere interlopers. In 1889, for instance, Samuel Pozzi described in detail the symptoms of a patient, Adele H.: regular discharge of blood, pains in the kidneys, very strong migraines, and breasts that became more voluminous and more sensitive during these periods.[191] All fairly typical late nineteenth-century signs of menstruation. Yet Pozzi called these phenomena "pseudo-menses."[192] Why? Because they appeared in an individual Pozzi was convinced was a man. In any woman these signs would have been interpreted as clear signs of menstruation, but appearing as they did in a person deemed medically male, they took on the status of merely apparent, even deceptive signs. The ultimate diagnosis of true sex was therefore reflected back onto sex signs, as traits that matched the "true" sex were labeled "real," and all the mismatched merely "apparent." Thus while a vulva might look quite the same in a true woman as it did in a masked man, the medical man insisted that in the true woman it constituted a "veritable vulva," and in the masked man only a "pseudo vulva." This again reinforced the hopeful belief that there really could be no true hermaphroditism in humans, no truly profound mixing of the male or female—since all anomalous signs were only "apparent" or "false." In this way, medical and scientific men in France and Britain again enlisted hermaphrodites' bodies to help shore up the sex boundaries they seemed at first to challenge.

ONE BODY, one sex: this was the magnet rule around which the rhetorical, conceptual, and physical treatment of hermaphroditism hovered. It is truly remarkable that in the midst of all the swarming doubt and sexual slippage, medical men—even those intimately familiar with hermaphroditism—managed to maintain the unflagging belief that there were two well-delineated sexes. Yet as we have seen, paradoxically, the whole medical approach to hermaphroditism was imbued with the assumption that there did exist two distinct sexes and only two sexes, and that, accordingly, each body ought to be limited to one, in theory and in practice.

CHAPTER 4

Hermaphrodites in Love

Lack one lacks both . . . and the unseen is proved
  by the seen
Till that becomes unseen and receives proof in its turn.

Showing the best and dividing it from the worst,
  age vexes age,
Knowing the perfect fitness and equanimity of things, while
  they discuss I am silent, and go bathe and admire myself.

—from Walt Whitman, "Song of Myself"

IN THE SUMMER OF 1892, one not-so-ladylike individual by the name of Louise-Julia-Anna presented herself to Dr. François Guermonprez of Lille. She had been referred to him by his colleague, Dr. Reumeaux of Dunkirk, with "no other information than this: 'subject interesting from the psychological point of view.'"[1] Though her face had been shaven clean just before their encounter and though she came wearing a lady's dress, corset, gloves, and hat (see Figure 13), Louise-Julia-Anna still struck Guermonprez as a rather poor specimen of a woman:

Her outfit is badly adjusted, lacking in grace and lightness . . . her brooch is placed poorly to the side; her girdle goes more to one side than the other; the flowers and the ribbons of her hat are disposed without taste and the entire ensemble bespeaks a sort of negligence, which is not the consequence of bad intentions, but which results mainly from absence of good taste.[2]

Still, Louise-Julia-Anna's unrefined taste in dress was not what made her so "psychologically interesting" to the doctors. It was, rather, her taste in lovers: she desired only men. Indeed, she had had sexual intercourse with more than one man, but never a woman.

What did the doctors find so remarkable about this? Although Louise-Julia-Anna had been raised as a girl, and although she appeared, and believed herself, to be female, Reumeaux and Guermonprez were convinced that she was a man. The patient had first come to Reumeaux seeking treatment for an inguinal hernia. But during a preliminary examination, the first doctor had discovered, much to his surprise, testicles, as well as what looked like a small penis:

> Stupefied, [Reumeaux] interrogated [Louise-Julia-Anna] with prudence. This person thought herself a woman; she had had relations with men and showed no attraction toward persons of the feminine sex. There was nonetheless no doubt anatomically about the masculine sex of the subject [because she had testicles]; but the questions were rephrased in vain: it always resulted in receiving responses which revealed the exclusive penchants of the feminine sex.[3]

If, the doctors wondered, Louise-Julia-Anna was a man, as they were sure, why did she desire other men, and only men? In Guermonprez and Reumeaux's eyes, the combination of the subject's "male" anatomical sex ("male" because of her testicles) and her "feminine" desires ("feminine" because directed toward men) constituted a "bizarre contradiction between the anatomical worth of the subject and the psychic characteristics of her sexual tendencies!"[4] This "man" had "womanly" desires, so, the doctors figured, there must be something wrong psychologically with Louise-Julia-Anna. Guermonprez and Reumeaux concluded that their patient was "truly a teratological being morally as well as physically."[5]

Guermonprez did not spare Louise-Julia-Anna the full brunt of his opinion: he told her she was a man, and instructed her to stop pretending otherwise. As might have been expected, "the revelation of the masculine nature of [her] sex troubled [the patient] profoundly." Guermonprez triumphantly noted, however, that there was "not a tear, not a sigh, not the least vestige of an attack of nerves! There was nothing of that pro-

Figure 13 Louise-Julia-Anna.

found distress of a *true* woman found in the presence of an event which reverses her life all at once."[6] The fact that Louise-Julia-Anna took the life-altering news with "a firmness thoroughly virile" convinced Guermonprez that his diagnosis was right: she must indeed be a "true male," a subject of masculine pseudohermaphroditism. Her sexual desires for men were therefore, in his eyes, thoroughly inappropriate. Three feminine names, coifed hair, a dress, a corset, sexual desires for men—all these supposedly "feminine" traits were not enough to make Louise-Julia-Anna a woman as far as the doctors were concerned. Her testes defined her as a man, and her doctors thereby demanded "manhood" from her—bravery in the face of terrifying revelations and a "masculine" desire for women even after a long life of "mistaken" womanhood.

AS NOTED in the last chapter, today distinctions are typically drawn between sex (which is considered an anatomical category), gender (which is thought of as a category of self- and/or social identification), and sexuality (a term used to refer to sexual desires and/or acts). At least until the late nineteenth century, however, such divisions did not exist. The characteristics we conventionally associate with these categories were believed to belong naturally together. The case of Louise-Julia-Anna illustrates this point. Guermonprez and Reumeaux assumed that testicles (a matter of sex to us) must entail bravery (a matter of gender to us) and desire for women (a matter of sexuality to us). Indeed, Guermonprez and Reumeaux presumed the sex-sexuality tie was so tight that, even if Louise-Julia-Anna did not know she was a man, her body would know it, and her internal masculine anatomy would drive her to desire women.[7] They therefore had expected this "masked man" to confess to loving women, not men. This chapter focuses on that one major assumption that framed and governed the biomedical treatment of hermaphrodites, namely, the assumption that true males would naturally desire only females and that true females would naturally desire only males.

Post-Darwinian French and British thinkers often declared that sexuality—specifically heterosexuality—was a natural outcome of the evolution of the two supposedly distinct sexes.[8] Just as "natural" sex differences were traced back to the primordial slime, so was the "nature" of heterosexuality. In 1894, for instance, Charles Debierre, professor of anatomy at the Medical Faculty of Lille and a man with a lively interest in all beings hermaphroditic, asked rhetorically in an article of the same name, "Why do males and females exist in the organic world?" His answer: to rejuvenate the species. The sexual encounter of the male and female would, literally and figuratively, bring new life to a species. Heterosexuality seemed a necessary aspect of survival. "Love is not the result of supernatural forces," Debierre declared, "but has, indeed, a very humble origin, being simply the result of the evolutive forces of living matter."[9] Heterosexual attraction and conjugation consequently stood not as issues of mere romance. Heterosexual pairing seemed instead a dictum of nature, the natural path of males and females, the very reason for their existences and differences. True males and true females therefore—even if masked by spurious hermaphroditism—were expected to follow the nature of their "true" sexes and to pair only heterosexually. And at this extended historical moment when "sex" had been reduced to gonadal tissue, it was fairly easy to figure out what heterosexuality was. In other words, given that sex had come to be equated with the gonads and only the gonads, one could fairly easily say medically what was a heterosexual encounter and what a homosexual one—even if, as we shall see, the consequences of this line of thinking were, especially in the realm of hermaphrodites, absolutely fantastic.

Investigations of Hermaphrodites in Love

It is important to note that not all cases of "mistaken" sex flew in the face of the assumption of automatic heterosexuality as the case of Louise-Julia-Anna seemed to do. French and British colleagues of Guermonprez and Reumeaux documented a number of instances in which pseudohermaphrodites had (unlike Louise-Julia-Anna) the sorts of desires the medical men expected from their "true" sex, even though the pseudohermaphrodites apparently did not consciously identify with that "true" sex. Thus it did not surprise Samuel Pozzi to discover that Louise Bavet, apparently a woman, "has a very pronounced taste for women, and men inspire no desire in her," or that "in her dreams, it is always women whom she holds in her arms." It was no surprise because "Louise B[avet] has testicles, thus Louise B[avet] is a man."[10] Men—even anatomically masked men—were supposed to naturally love women.

So pervasive was the sex-sexuality assumption that in a number of instances, the "true" sex of apparently female horses, goats, and people came to be doubted specifically because of unexpected "masculine" sexual behavior. Sometimes the suspicion that "strange" sexuality could signal mistaken sex paid off. In 1897, for instance, a Scottish filly fell under suspicion "because this seeming filly was quite vicious with other female equines, teasing and tormenting them like any entire colt or stallion. In fact, as the season came in she got quite unmanageable with other females in the fields."[11] A summoned veterinarian made a thorough investigation and discovered two testicles and a penis. He castrated the horse, removing the testicles and the "erratic penis" and was "very glad to say that the filly [no longer the 'seeming filly'] . . . made a splendid recovery in every way."[12] The pursuit of females ceased. Conversely, in a human case, the sudden descent of testicles in an apparent woman was considered responsible for the abrupt turn of her sexual interests from men toward women.[13] Such cases as these were quite easily interpreted in such a way as to support the sex-sexuality paradigm.

In France, medical men were especially persistent in their attempts to ferret out information about hermaphroditic subjects' sexual pursuits. In the late 1880s two young Parisian surgeons who performed a postmortem on the body of Marie, a person thought by her companions to be a woman, discovered that she had testicles and no ovaries. Unable to interrogate their now-dead subject about her sexual exploits, the surgeons instead sought out Marie's concierge. The concierge reported, to the satisfaction of the surgeons, that Marie manifested desires "inappropriate" to a woman: when the concierge's husband went away, Marie pursued the concierge with "caresses and lewd touching," relations which the concierge found *"ridiculous between women."*[14] (Medical men were by no means alone in their belief that women should naturally love only men, and vice versa.)

We can see even from these few vignettes that, just as the status of a vulva as "veritable" or "pseudo" depended on what was judged to be the "true sex" of the body in question, the status of a behavior or desire also depended on the perception of what was "real" and "true." So, for instance, Marie's desire seemed inappropriate to the concierge, who knew Marie in her lifetime, but to the doctors, who (literally) saw Marie quite differently after her death, Marie's sexual desires now seemed natural. Behavior apparently "perverse" became understandable when the internal "truth" was revealed, and inversely, a new gonadal "truth" could uncover unsuspected "perversity," as in Louise-Julia-Anna's case.

The result of this situation was that hermaphrodites often possessed two different identities, one personally and socially, another medically and scientifically. To scientifically minded French doctors, the "truth" was always internal, even when invisible and seemingly unmanifested. In fact, Guermonprez refused to treat Louise-Julia-Anna's hernia (shown in Figure 14) because she refused to live by his truth; she made it fairly clear she intended to go on living as a woman, and she only wanted her hernia fixed because it made relations with men difficult. As far as Guermonprez was concerned, "the operation requested appeared to me to become, no longer an *operation of complaisance,* but rather an *operation of complicity; something like a profanity of the art* [of surgery]. This is why I have refused!" He would not participate in enabling a "man" to have sex with "other" men.[15]

Compared with their French counterparts, British medical men were much less interested in inquiring into hermaphroditic patients' sexual

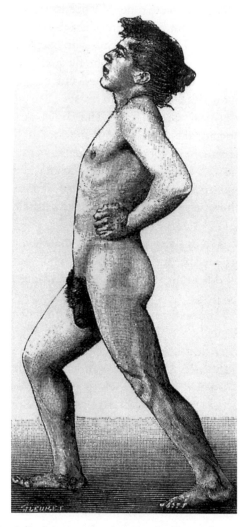

Figure 14 Louise-Julia-Anna, nude, showing the hernia hanging down between her legs.

penchants and exploits—or at least less inclined to report their findings in this realm. Nonetheless, they did share with their French colleagues the assumption that gonads determined sex and sexuality and so they accepted the type of sexual predilections as a strong indication of the "true" sex of a individual. For example, one English physician took as powerful evidence for maleness the fact that C.D., a kitchen maid, confessed to "male" sexual desires: "[T]he sexual and sensual indications of this individual are for a female, and not for a male; . . . a longing for connexion with a girl has been experienced, though never indulged."[16]

On both sides of the Channel, male sexuality was thought naturally, frightfully insidious, and the possibility (and occasional incidence) of masked males placed among unsuspecting females caused medical men terrific anxiety. Indeed, the pernicious character of male sexuality constituted a major reason for raising doubtful children as males: if a masked male were raised among females, the logic went, his sexuality was bound to erupt, forcing him to behave "like a wolf in a sheepfold."[17] Accordingly Lawson Tait instructed practitioners faced with doubtful children:

> If . . . no genital orifice can be discovered, let the patient be considered as a male, for if brought up amongst males but little harm can come to him. If, however, an individual were brought up amongst girls who turned out to be a semi-competent male, no end of mischief might accrue, as is amply proved in the case of Madelaine Mugnoz, the nun of Ubeda, who suffered death for rape.[18]

When Tait said "little harm can come" to such a patient, he meant that no vaginal intercourse could occur, because there was no vagina. Tait's warning was positively clinical compared with Neugebauer's exposition of the worst-case scenario. The latter described in lurid detail

> the supposed young girl of 16 [in reality a masked man] . . . her sexual instincts aroused by reading romances; sharing the dormitory of her fellow pupils, she watches every night and morning these maidens at their toilet, and without any restriction has opportunities of learning to admire the bodily charms of persons of the opposite sex.[19]

Such circumstances, Neugebauer presumed, would naturally serve to "awake[n] in him, however innocent hitherto in regard to sexual matters, desires that might easily lead him to abuse the situation, in a way which might be followed by consequences even more serious perhaps for the companions sacrificed to his masculine desire than for himself."[20] It is worth noting that Neugebauer so firmly associated naturally strong desires for females with masculinity that he switched the gender of his pronouns midway through his fantasy. The hermaphrodite began as a "she" but became a "he" as powerful desires for "his" girl classmates emerged. At least when it came to male sexuality, it seemed that true sex would find its way out, no matter how thick the veneer of abnormality. One could hardly expect any "real" man, swimming in a sea of feminine bodies, to suppress his desires. Masked male sex could only lead to the raining down of "demoralization and scandal."[21]

In France, medical men warned that hermaphrodites who looked like men externally but had ovaries internally were likely to seduce unwitting men. In 1885, for example, one French writer hysterically warned about the two-way perils of mistaken sex:

> Grave social disorders result from these mistakes. Put one or the other of these mal-sexed individuals into religious orders or a teaching position, and morality will be gravely compromised. If a man-woman [that is, a mannish-looking woman] were admitted to the seminary, what would become of the young Levites at her contact . . . ? Still much more dangerous would be a woman-man [that is, a womanish-looking man] like Brade [a male who mistakenly landed in a nunnery]; there too the fire of the convent consumed the nuns! It would be worse still in the schools, in the boarding schools, the grammar schools. What would happen in the barracks if the preliminary examination of recruitment did not assure today against similar mistakes?[22]

A hundred years ago, hermaphrodites in the military posed a frightening risk, given that their "natural" sexuality seemed irrepressible and uncontrollable. Not only did they threaten to seduce other soldiers, but sometimes by doing so they even "brought a child into the world in open camp!"[23] Faced with a pseudohermaphrodite, one could not be assured of a clear separation of the sexes, and therefore one could not ensure against mixing "true" men and women where they ought not be mixed lest their supposed "natural" sexual drives toward each other force them to couple.[24]

Yet, conversely, some medical experts also feared pseudohermaphrodites (and those people who intentionally cross-dressed) because they might, by appearing to belong to the other sex, tempt unwitting souls into same-sex coupling.[25] Émile Laurent, a French doctor who readily immersed himself in the exploration of the anatomy and physiology of sexual peculiarities, insisted that "to place a young man with the breasts and grace of a woman in a barracks would be practically to encourage pederasty."[26] Laurent bemoaned how often "accidental hermaphrodites" were formed when young boys and men, "already feminine in their form and their allure become more so by education" (that is, through "indoctrination into pederasty") until "their souls become feminine, too."[27] In England, a 1901 editorial in the *Lancet* similarly expressed profound contempt for "the vilest of the vile," namely those "men of effeminate appearance dressed as women" who enticed men seeking women prostitutes into unwitting same-sex intercourse.[28] This fear, that "normal" males might be tempted by feminine-looking men into sexual relations, directly contradicted the running assumption that sexual desire would always be determined by one's "true" anatomical sex, no matter how masked that sex. Obviously a general fear underlay all these anxious pronouncements, namely a fear of all sexual encounters that did not conform to particular kinds of socially sanctioned heterosexual relations.

The Problem of "Same"-Sex Marriages and Surgical Interventions

In the last decades of the nineteenth-century, many French and British biomedical experts expressed profound horror at the very real possibility that two persons could enter into marriage without anyone, even the spouses, realizing that both spouses were "really" (meaning gonadally) the same sex. France was by the late nineteenth century a nation burdened with a population crisis, one "believed to be as much a biological and moral crisis as one of voluntary reproductive parsimony," all manifested in "an extraordinary slackening of their birth rate."[29] French medical men therefore held a deep concern for productive marriages, and were particularly troubled by "mistaken sex" marriages which, though proper in the eyes of the partners and their acquaintances, were doomed to be sterile and were, in the eyes of the doctors, fundamentally "malformed." Indeed, at least one anxious French doctor insisted that every sterile marriage be considered a possible case of mistaken sex, while still another admonished his colleagues:

> The question [of mistaken sex] which occupies us has only a relative gravity if it concerns only contested heritages, electoral rights, or military service; but it is entirely otherwise when marriage intervenes. One can then be found in the presence of monstrous alliances, and see, for example, two men or two women united together, by a mistake which engenders social disorders, causes scandalous divorce, or creates some wretchedly equivocal situations.[30]

Time and again doctors' discoveries of an unexpected "true" sex arrived at a point when "the law had already sanctioned a union against nature."[31] One French physician warned that these "monstrous marriages" (like the one of the couple shown in Figure 15, in which the wife was a male pseudohermaphrodite) would only wind up disastrously unhappy, because "the sexual appetite of the individuals cannot be satiated"—as if a relationship between two men, because "unnatural," could never be sexually satisfying, even if everyone believed one of the two was a woman.[32]

Yet there was much evidence to the contrary. Sophie V. and her husband were not alone in their "same sex" marital happiness. Neugebauer's ever-growing collection of "marriages made between persons of the same sex" made it absolutely clear that pseudohermaphrodites often married persons of the same gonadal sex, and that, furthermore, in many cases they remained so married until or even after a medical expert or an unexpected happenstance of physiology (like the onset of menstruation in a supposed man) revealed a long-standing "error of nomenclature."[33]

Some hermaphrodites' anatomy allowed for relations resembling vaginal intercourse, so that, as far as the partners were concerned, nothing about their relationship was bothersome. Of course, the genitalia of a hermaphrodite could be such that it was impossible to engage in intercourse without serious pain or difficulty, a condition which often—as in the case of Sophie V.—precipitated a medical examination and the declaration (to the patient or privately to colleagues) that one spouse was not the sex that everyone had assumed.

In France, sex-change procedures were often quite public.[34] Errors of sex when discovered in adults were frequently announced to the patient and sometimes the larger community and were legally amended.[35] Recall, for example, the 1885 case of Joseph-Marie X., formerly known as Marie

Figure 15 Photograph of a married couple believed by medical men to both be gonadally male. Caption to the original reads: "A.S. [shown on the left], married for sixteen years, male hypospade. Error of sex."

X., in which the tribunal of Château-Gontier called in three medical experts to pronounce their opinion about X.'s sex. (X. sought to have his legal status formally changed to conform to what he thought was his true sex, that is, male.) The case of Joseph-Marie X. seems to have been as much lacking in consideration for the privacy of the plaintiff as the case of Barbin. It is also worth noting here that in Joseph-Marie's case, the panel of experts agreed that the plaintiff was a male in part because "on various occasions, she claims to have had some erotic dreams pushing her to be drawn to persons of the feminine sex."[36] They assumed this meant that "she" had testicles, and the legal sex change was granted. Joseph-Marie's sexuality was thereby normalized.

In Britain, meanwhile, matters of mistaken sex and sexuality generally proceeded along more confidential lines.[37] Indeed, British surgeons also seem to have made something of a habit of removing apparent-females' testicles at least in part out of fear of the irrepressible desires of the male gonads. As we saw in the last chapter, upon discovering testicles in nine-year-old Christina, a Glasgow surgeon decided to excise them immediately, because he firmly believed that "it would be a singularly unfortunate thing for a person with all the outward appearance of a female to develop into a man as to internal organs and feelings."[38] A short time later, the surgeon presented the case history, Christina herself, and her testicles to the local Medico-Chirurgical Society and was gratified to hear his colleagues assent to "the propriety of the operation."[39]

The treatment of Christina's condition did not form an isolated history in Britain.[40] Recall the widow from Chapter 3 who came to Middlesex Hospital and complained to the surgeon Andrew Clark of a painful swelling in the groin. Clark suspected testicles even before operating, and when he opened the widow's labia he found his diagnosis vindicated. There was "no doubt about the nature of the organs" he subsequently removed. They were testicles. Still, in the end, Clark "did not think it necessary or even fair to inform her of what we had discovered, and when she left the hospital she believed, as far as I am aware, that she had been suffering from an ordinary rupture which had been cured."[41] Apparently in this 1898 case Clark decided that, since the patient was by then relatively mature and left widowed by her husband of sixteen years, it would not matter if she went on in a female role.

Most likely, like his compatriot colleagues, Clark assumed that since he had removed her testicles, he had in fact "unsexed" her, so that she was then, if not a woman, at least not a man and free of any latent, dangerous, "male" desires. G. R. Green of Ripon decided in favor of advising his patient that she possessed two testicles. Twenty-four-year-old A.H. had come to the surgeon in January 1897 to complain "of feeling weak and short of breath" and to obtain his opinion as to why she never menstruated.[42] His answer to the latter question must have surprised her. She looked very much like a woman (see Figures 16 and 17), she was employed as a woman, and she thought of herself as a woman. Inevitably "[t]he question now arose as to what should be done, and as the patient in mind and habit is more a woman than a man, and it is illegal for him to remain as he is in female attire, he expressed a desire to have the testicles removed and continue a woman."[43] Green shared A.H.'s opinion

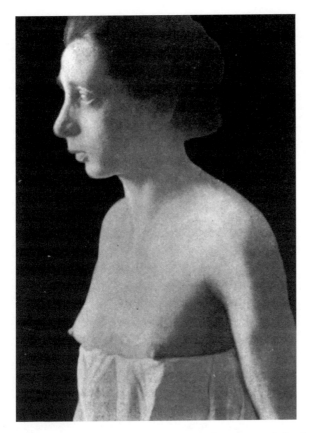

Figure 16 Photograph of the upper body of A.H., who requested that her testicles be removed by the surgeon G. R. Green so she could continue living as a woman; he obliged. Caption to the original reads: "To show feminine type."

that this was "the best solution of the difficulty," and accordingly he performed a double castration at the Ripon Cottage Hospital. A.H. then, in Green's words, "*continue[d]* a woman," that is, "he [*sic*] made a good recovery, and having been unsexed, has now returned to domestic service as a housemaid."[44] Paradoxically the removal of testicles in these sorts of cases in Britain actually demonstrates how important they were considered to be—important enough to warrant prompt removal in any apparent girl or woman who was to "remain" so. It seems that British surgeons generally preferred these "desexing" procedures to the French alternative, that is, the public admission of a case of hermaphroditism or longstanding mistaken sex. (This also seems to be in keeping with the relative

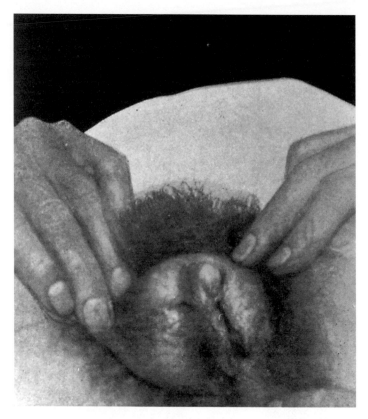

Figure 17 Photograph of the genitals of A.H., held open to show the phallus. Caption to the original reads: "Showing penis, groove, and urethral orifice."

reluctance on the part of British physicians and surgeons to discuss at length their patients' carnal desires and experiences.)

Occasionally, a French surgeon allowed an apparently same-sex union to continue rather than taking on the "ungrateful role" of destroying a happy marriage by announcing mistaken or doubtful sex.[45] Take, for instance, the case of Madame X of Angers. Though she had received several proposals of marriage when a young woman, her parents "did not judge it honest to consent to a marriage destined to remain sterile . . . since the young girl never had the menses."[46] But when Madame X reached the age of forty-four in 1898, a sixty-year-old widower who already had several children and grandchildren found her sterility a fine selling point and asked this "woman" to marry him.[47] Try as he might, Monsieur X could not penetrate his new wife to the depth he desired, and his attempts caused her some pain. The two, however, did not find the situation entirely unsatisfying: "[S]he felt voluptuous sensations when ejaculation was produced in her husband, [and] these sensations reached their climax in the form of rhythmic spasms accompanied by a shaking of the whole body and an emission of sticky liquid in the area of the vulva."[48] After Madame X suffered a fall, she discovered two new "tumors" in her labia majora, and this combined with the painful intercourse and several other complaints (dizziness, giddiness, vague pains) convinced her to confer with a doctor in 1899, although her friends assured her the symptoms merely signaled menopause.

During her consultation with Dr. Raoul Blondel in Paris, Madame X took the opportunity to ask whether the surgeon could not do something to enable "more complete relations with her husband,"[49] for she "loved her husband tenderly."[50] A rather troubling question, for Blondel had already decided the "tumors" in the patient's labia signified not menopause, but manhood: they appeared to be testicles. Other signs also pointed toward mistaken sex, including her flat-chestedness, sizable "clitoris," and failure to menstruate. At the same time, her "voice, pelvis, character, and inclinations" were all rather "feminine."[51] True enough, she managed some sexual relations via her short, blind, "pseudo-vagina," but as far as Blondel was concerned, "from the social point of view, it [was] evident that the marriage of this unfortunate [was] null" by virtue of his belief that her "true" sex was male. Two men could not be legally married. Nevertheless, Blondel and a colleague he consulted decided that their nagging "doubt should be interpreted in favor of the subject's intentions." Blondel therefore offered Madame X the help she sought: he "decided to section her hymen and elongate her pseudo-vagina by an incision at the base of its cul-de-sac . . . [and] by means of grafts."[52] She agreed to undergo the surgery (apparently unaware of his chief diagnosis), and made an appointment for the procedure, though for reasons unknown she did not keep it.

Similarly, although he did not perform any surgery designed to "better" his patient's sex, Dr. Debout of Rouen did, like Blondel, let a mistaken patient go without insisting on a rectification. Debout found himself "in the presence not of a young girl [as it first appeared], but of a hypospadic boy" who came to him for a painful hernia not realizing his "true" state.[53] Two "colleagues and friends" of Debout confirmed his diagnosis, yet Debout apparently did not inform the patient of what he had found. Instead he made the male hypospade a custom bandage to relieve her pain so that she could return to her occupation of weaver.[54]

Nonetheless, in France, this sort of aiding and abetting of "mistaken" sexes and "same"-sex relations did not go unchallenged as it seems to have in Britain. One French physician, Dr. Xavier Delore, sharply reprimanded Blondel and all other medical men who would let continue, and even promote, the possibility of "essentially immoral" unions of two true males or two true females, no matter how honorable the intentions of the parties.[55] In Delore's opinion, the similarity of the partners' sexes constituted "a most profound vice in the origin of the marriage and consequently in its validity."[56] The practitioner faced with such a discovery must be "guided by reason and sentiments of high morality"[57] and see to it that the marriage be dissolved, not remedied, lest a doctor like Blondel lend himself "to a conjugal onanism of a new genre!"[58] Similarly, a pair of French doctors criticized the English surgeon Green for removing the housemaid A.H.'s testicles. They admonished, "It is necessary *to rigorously abstain from all operations which are inverse in sense to the [true] sex.*"[59] Clearly on both sides of the Channel it had become the medical man's duty to police the "natural" law of heterosexuality, although some differences of opinion existed as to which tactics of enforcement should be deployed.

"Homosexual" Hermaphrodites

In retrospect, it is remarkable that, despite so many well-documented instances in which people of a hidden "true" sex expressed the desires proper to their "apparent" rather than their "true" sex, French and British medical men continued to labor under the assumption that testicles naturally meant desire for women, and ovaries naturally meant desire for men.[60] As we have seen, in a sizable number of cases of pseudo-hermaphroditism, ambiguous people displayed desires unequivocally for beings of the same gonadal sex as themselves. These apparent anomalies to the sex-sexuality paradigm obviously required explanation in ways that true-male/true-female coupling did not. How were they explained, then?

In France, where an elaborate psychology of sex was under development, some experts on hermaphroditism suggested that "mistaken" upbringing could account for why a male pseudohermaphrodite might seem to desire only "other" males. One, for instance, posited that "education can . . . push the sexual instinct in an opposite direction of that which it should be, and thus produce a sort of sexual inversion."[61] Still, some French authors preferred the explanation that "confused" sexual desires (desires for the "wrong" sex) were the inevitable result of confused sexual anatomy, that they were of one pathology: strange anatomy, strange sexuality. Louis Blanc told lay readers of his picture-packed book on monstrosities, "morality itself is often modified by this perversion of the genital apparatus."[62] But in France another increasingly favored explanation also arose, namely, the idea that people like Louise-Julia-Anna were congenitally homosexual hermaphrodites—that they suffered not only from congenitally ambiguous sex, but coincidentally from congenitally "inverted" (homosexual) sexuality. They loved the "wrong" sex because they were destined to do so regardless of their anatomical ambiguity.

We should pause here to note several basic points in the history of scientific treatments of homosexuality and heterosexuality. Historians of sexuality have convincingly documented that homosexuals—and by consequence, heterosexuals, too—as we generally now know them came into being only recently, specifically in late nineteenth-century Europe. These historical claims are not at all meant to imply that people in other places and times never experienced what we now think of as homosexual and heterosexual desires, nor do they mean that what are now thought of as homosexual and heterosexual acts were not performed with regularity before this time. But it was only a little more than a hundred years ago that the "homosexual" rather suddenly became, as Michel Foucault put it, "a personage, a past, a case history, and a childhood, in addition to being a type of life, a life form, and a morphology, with an indiscreet anatomy and possibly a mysterious physiology."[63] Relatively suddenly in the last decades of the nineteenth century, the "sexual invert" became established as a distinct human type, and with that emergence of a distinctive "homosexual" identity there also grew up over time a definitive "heterosexual" identity. The twentieth-century West then came to own a grand—and seemingly timeless—division of sexual types into the heterosexual and the homosexual.[64]

This claim may seem to contradict the position, stated earlier, that issues of what we call sexuality were not really disassociated in the late nineteenth century from issues of sex. It seems that, if the types of "heterosexual" and "homosexual" emerged, people must have started to think about sexuality as a separate issue from sex. How else would homosexuality be possible, since homosexuality requires that it be conceivable that a male could have supposedly "feminine" desires? But sexuality was—and is—still very much understood in relation to sex. Indeed, the very classifications of "homosexual" and "heterosexual" have, of course, never ceased to contain within them critical references to sex and in fact they require sex: one category specifically indicates that the partners are of the same sex, the other that the partners are of different sexes. Sexuality did not and has not grown to be such a separate concept from sex that it does not constantly still make reference to anatomy. Even though homosexuals and heterosexuals as definite types became conceivable in the late nineteenth century, sex and sexuality did not grow apart to become entirely separate issues, and in fact, judgment about the normality or abnormality of any given sexual relation continued typically to rest on the assumption that the only normal, natural form of sexuality was heterosexuality—that a female loving a female or a male loving a male was so very startling that it deserved extensive scientific research and classification and possibly even medical treatment.

Thus during the last two decades of the nineteenth century, a contingent of French medical doctors, many of whom were also interested in hermaphroditism, labored to enunciate a science of "inverted" sexuality, that is, a science of the phenomena in which women desired sexual relations with other women, and men with other men. In 1882 Jean-Martin Charcot and Valentin Magnan presented the first substantive French study of "sexual inversion," an account that was considered definitive for many decades.[65] Of special interest to all who participated in these discussions was the question of whether sexual inversion could be congenital. Most concluded it could be.

This conclusion came directly to affect the French conception of hermaphroditism. By February of 1911, the French surgeon and hermaphrodite expert Samuel Pozzi had teamed up with the same Valentin Magnan to present to the Academy of Medicine in Paris a full-blown study of "the inversion of the genital sense" (homosexuality) in hermaphrodites.[66] Pozzi and Magnan's 1911 presentation began with the history of one recently discovered "feminine hermaphrodite," specifically a person who looked like, had been raised, and considered himself a man, but was confirmed to be female internally. In fact, the patient had come to medical attention because of an abdominal tumor that turned out to be ovarian in nature. Neither the patient nor his wife, who loved each other very much, suspected anything was amiss with regard to the husband's sex. "To himself, he is a man, and the same conviction exists in his wife; she has a husband, a true husband."[67] Yet the doctors were shocked—not so much by the remarkable contradiction between the internal and external anatomy of this subject as by the contradiction between his true "female" sex and his "masculine" sexuality. Magnan and Pozzi concluded this patient's "homosexual" desires could only be explained as "a matter of the brain of a man in the body of a woman," or, to use the by then standard language of sexual psychology, a matter of "complete inversion of the genital sense."[68]

In short, Pozzi and Magnan did not suggest that this patient's supposed "sexual inversion" might be due to a masculine upbringing and identity. Nor did they suggest that the inverted sexuality might be due to any dysfunctionality on the part of the tumorous ovaries.[69] Instead Pozzi and Magnan classed this patient among the many anatomically normal "homosexuals" of whom they knew, and considered the new case simply another of congenital inversion of the genital sense (congenital homosexuality) in a person who also happened to be a hermaphrodite. In fact, they drew an explicit comparison between this case and one of "innate inversion of the genital sense in a[n anatomically] normal subject," a military officer.[70] Both subjects were, in Pozzi and Magnan's eyes, cases in which true males desired only true males.

"That which for the vulgar constitutes the principal interest in these malformations [of anatomical hermaphroditism]," Pozzi reported in his remarks, "is the crude idea of a double sexuality physiologically in these exceptional subjects."[71] But Pozzi objected to this notion. Drawing on his extensive familiarity with hermaphrodites—he had by then seen nine cases of anatomical hermaphroditism personally—the renowned surgeon insisted that "reality does not correspond to this hypothesis. Far from being bisexual from the point of view of the genital sense, most pseudohermaphrodites could rather be qualified as being *asexed* or at most *oligosexed*."[72] ("Oligo" means reduced or deficient.) Still, Pozzi noted, hermaphrodites *could* exhibit sexual desire, and he argued that this desire should be classified according to the true (gonadal) sex of the individual as it was in anatomically normal subjects. Just as homosexual partners were to be recognized by and labeled according to the similarity of their gonads and heterosexual partners by the dissimilarity of their gonads, so was the sexuality of hermaphrodites to be judged and classified. "Homosexual" hermaphrodites would be those who loved people of the same gonadal sex.

Pozzi therefore recommended labeling the sexuality of each hermaphroditic subject as follows:

I.—*Asexed* or *oligosexed.* Subjects indifferent or nearly indifferent from the sexual point of view.
II.—*Homosexed* or *inverted.* Among these, we can admit a subdivision: In one category, the inversion appears very much to be a secondary effect of causes acting *artificially,* if we may say it this way,

on the mentality and the habits of the subject [as through "mistaken" sexual education]. In another category, it seems that the inversion was original or *innate*.

III.—*Heterosexed* or individuals having the sexual appetite directed toward women if they have testicles, toward men if they have ovaries. (It might be preferable . . . to call them *orthosexed,* that is to say, sexed in the normal direction.)[73]

For Pozzi and Magnan and their colleagues, just as "true sex" was understood to be gonad dependent, so was "true sexuality." Thus the identity of one as a sexual being—as a being with sexual desires—was to be judged according to that desire's relation to the gonads, not according to its relation to the individual's social or personal identity. And so there came to be homosexual (pseudo) hermaphrodites.

In 1911, the same year that Pozzi and Magnan offered their fully articulated method for classifying hermaphrodites according to the "truth" of the hermaphrodites' sexual desires, two of their French colleagues applied just such a taxonomic system. Mademoiselle L.S., aged twenty, first presented herself at the Hospital Beaujon in 1909. She was about to be married and wanted some bothersome tumors in her labia majora removed before her wedding day. At first the men who examined her did not question her sex. In fact, L.S. was considered such a lovely specimen of feminine grace that she had been a fashion model in Paris. (Perhaps this is why she was so posed in the photographs that accompanied the medical report of her case; see Figures 18 and 19.) But a biopsy of the tumors showed them to be testicles, and L.S. was accordingly labeled by her doctors a true male, that is, a masculine pseudohermaphrodite. The knowledge that L.S. "did not hide her taste for men" and the additional information that her "genital aspirations" had always been "feminine" led her examiners to feel forced to conclude that "L.S. is frankly homosexual."[74] L.S., the beautiful fashion model, was a homosexual because she loved men and only men. The truth of her identity lay in her gonads.

"Hermaphroditic" Homosexuals

In the late nineteenth century, meanwhile, British medicine did not experience the burgeoning interest in "sexual inversion" that French medicine did.[75] In theory, as in practice, the British were inclined to deal with

Figure 18 Photograph showing L.S., the Parisian model, designated "frankly homo-sexual" by Tuffier and Lapointe because sexually she desired men and not women. L.S. had testicles and today would likely be classed as a case of androgen insensitivity syndrome (AIS).

anatomy rather than psychology, as when British surgeons "desexed" apparently-female patients. (We may wonder, had Louise-Julia-Anna been a British subject, whether, rather than being labeled "psychologically interesting" and told she was a man, she might have instead had her testicles removed and her hernia repaired.)

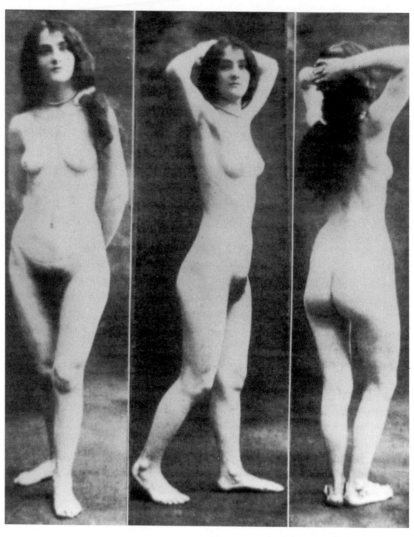

Figure 19 Three nude shots of L.S. The original caption asks the reader to note "the virilism of the inferior members [i.e., the lower limbs], in particular the feet."

Although medical approaches in France and Britain toward women with testicles may have differed considerably, at their foundation they shared the same basic assumption: that it did not make sense for people with testicles to have the desires "naturally" and exclusively appropriate to women, nor did it make sense for people with ovaries to have the desires that supposedly belonged to men.[76] Indeed, many medical observers wondered aloud whether a person who was the subject of "sexual inversion" might not also have something anatomically amiss. Voice was occasionally given to the floating suspicion that close examination of a sexually inverted person might turn up some anatomically hermaphroditic traits. So for instance in a 1896 "Note on Hermaphrodites" in his *Archives of Surgery,* Jonathan Hutchinson wondered "whether in . . . instances [of inverted sexuality] organic peculiarities which have been overlooked do yet really exist."[77] Perhaps those oddly "feminine" and "masculine" desires did sometimes have a physical basis, as they seemed to in the sexually "normal"?

When given the opportunity, those investigating sexual inversion often made thorough examinations of the bodies of subjects.[78] Case studies frequently contained remarks on the "femininity" or "masculinity" of the face, chest, skeleton, and genitalia. Reports abounded, for instance, of the prevalence of gynecomastia (enlargement and "feminization" of the breasts usually associated with undescended or atrophied testicles) in inverted men, even to the point where the male subject's nipples secreted milk.[79] Usually, as with pointed penises and infundibuliform anuses,[80] gynecomastia was figured to be a stigmata of pederasty, in this case a bodily metamorphosis that signaled prolonged indulgence in supposedly "feminine" forms of sexuality. Nonetheless, questionnaires designed to investigate the bodies, desires, and habits of the sexually inverted often contained questions about the condition of those physical traits commonly thought to be congenitally sex specific, such as the build of the body, the weight and shape of the bones, the density of body hair, and the type of "sex organs" including the genitalia and breasts.[81] Suspicions hovered that hermaphroditic bodies went with inverted desires.

In the first two decades of the twentieth century, those medical writers who gathered large amounts of data on "inverts" often noted, as did Richard von Krafft-Ebing in Austria and Havelock Ellis in England, that anatomical ambiguity was not always—or even usually—found in conjunction with sexual inversion. Collected evidence instead suggested that in most cases "the genitals are normally developed, the sexual glands perform their functions properly, and the sexual type is completely differentiated."[82] Robert Nye has noted that outside of France and its national sexual dilemmas, sexologists "could readily conceive of homosexual men who were masculine in every visible way and differed from other men only in their interior psychic constitutions."[83]

This image of the anatomically "normal" invert was not always the picture painted, however, and many sexologists inside and outside France still retained nagging suspicions that the bodies of at least some inverts leaned toward the hermaphroditic.[84] Ellis, for instance, noted:

> It seems to me, on a review of all the facts that have come under my observation, that while there is no *necessary* connection between infantilism [the persistence of childish features], feminism ["feminine" features in a man], and masculinism ["masculine" features in a woman], physical and psychic, on the one hand, and sexual inversion on the other, yet that there is a distinct tendency for the signs of the former group of abnormalities to occur with unusual frequency in inverts.[85]

Krafft-Ebing similarly suggested that while the genitals of "those individuals of contrary sexuality" were virtually always clearly differentiated into the normal male and female types, their "skeletal form, the[ir] features, voice, etc." might be such "that the individual approaches the opposite sex anthropologically."[86]

Moreover, while it is true that the men who wrote about inverted men often found it quite easy to imagine thoroughly "masculine" male inverts (anatomically speaking), by contrast their female subjects and archetypes of inversion were regularly also described as "masculine." In his *History of Similisexualism,* privately distributed in the early twentieth century, Xavier Mayne (the pseudonym of Edward Irenaeus Prime Stevenson) insisted the "Uranian, or Urning," that is, a man who desired men, "may be defined as a human being that is more or less perfectly, even distinctively, masculine in physique."[87] Mayne admitted that "Uraniads," that is, women who desired women, did not necessarily display anatomical ambiguity: "As is the Uranian continually in externals 'perfect man,' so also is the Uraniad, incessantly a 'perfect woman,' in her physical appearance, her manner, and all that is not intimately sexual. There is no necessary question of hermaphroditism . . . Often a perfect Uraniad is a veritable Venus, realizing the fullest feminine loveliness and grace." But Mayne continued:

> [T]here is to be admitted in the Uraniad [woman-loving woman] class a tendency toward imperfect sexual organization and functions; to divergences from the delicacy of the female anatomy. The Uranian [man-loving man] is likely to have nothing saliently feminine as to his

general physique and personality, and to possess most perfectly male organs. On the contrary, the Uraniad is often obviously "boyish" when a girl, has unfeminine proportions, bizarre muscular strength, and activity; . . . Also as she matures she frequently coarsens in body. When this occurs . . . we have the heavy-set, "mannish" woman, with a masculine walk and carriage, with a male timbre of voice . . . The Uraniad's [facial] features can be in due female proportion, but often of hirsute tendency, even to her showing a beard or moustache. Almost all "bearded women" are more or less Uraniad, and of "contrary" sexualism.[88]

Ellis similarly suggested that inverted women were often quite "feminine" in external appearance, but noted that *"truly"* inverted women—those who carried a "congenital predisposition" for inversion—were known often to betray "a more or less distinct trace of masculinity."[89]

Indeed, in a sense the congenital invert was always seen as a hermaphrodite of a sort, even if not anatomically one, as long as the desire for men was implied to be a "feminine" trait, and the desire for women a "masculine" trait. This lingering mentality revealed itself especially in the term "psycho-sexual hermaphrodite," a term which appeared in the last years of the nineteenth century. The "psycho-sexual hermaphrodite" was not a person with any anatomical peculiarity per se. Instead the label was used to denote an invert, or more commonly an individual in whom "there exists a sexual attraction to both sexes"—what we would call (with not much different connotation) a "bisexual."[90] The psycho-sexual hermaphrodite earned her/his title precisely from the idea that s/he had not a unique sort of desire, but a *doubly* natured desire, part "masculine" and part "feminine."

And, in fact, just as scientific and medical men expressed doubt that there could be more than a small smattering of *true* anatomical hermaphrodites among humans—so inconceivable was the simultaneity of male and female traits—even those sexologists who believed in a "congenital" form of inversion expressed doubts that *true* psycho-sexual hermaphrodites (bisexuals) could exist in any significant numbers. Ellis, for instance, suggested that in most supposed psycho-sexual hermaphrodites, either the homosexual or the heterosexual "instinct" was congenital, with the other added through circumstances of life.[91]

Writings on the subject of sexual inversion had always been laced with references to hermaphroditic states in part because many subjects themselves described the feeling of having some important characteristics of the "opposite" sex—characteristics that included (but were by no means limited to) seemingly incongruous "masculine" or "feminine" desires. For instance, in the 1860s–1870s, the would-be legislative reformer Karl Ulrichs, in describing the "third sex" in which he placed himself, had posited the existence of a "feminine soul confined by a masculine body."[92] Ulrichs, a "seminal figure in bringing about new thinking and research on same-sex love," fought to end the social and legal oppression of homosexuals by arguing that there existed a kind of homosexuality which was a natural and congenital variant of sexuality.[93] Although he did not imagine himself to be a physical hermaphrodite, Ulrichs did study reports of anatomical hermaphrodites and found in them a profound analogy to his own condition. He recalled the scientific explanation given for the origin of anatomical hermaphrodites: each embryo contained the elements of the male and the female, and the hermaphrodite constituted a being who, in development, or from a lack of it, did not see the usual suppression of one sex in favor of the other. Perhaps, Ulrichs suggested, each embryo also contained noncorporeal germs for both a "feminine" and a "masculine" sex drive. A Uranian (a man who desired men) would then be someone who had taken the masculine path in sex development, and the feminine path in sexuality development.[94]

Ulrichs viewed this sort of soul/body hermaphroditism as a variation on a theme of nature, just like the gradations in sexual anatomy manifested in the forms of anatomical males, females, and hermaphrodites. Following Ulrichs's lead in his search for a more tolerant explanation of homosexuality, Mayne made the homosexual/hermaphrodite analogy explicit, arguing that those possessed of a "similisexual [homosexual] instinct" formed "a series of originally intermediary sexes—the so called intersexual theory—rather than mere aberrations, degeneracies, psychic tangents, from the male and female."[95] The homosexual was, in Mayne's scheme, a natural psychological intergrade just as the hermaphrodite was a natural physical intergrade. Like Ulrichs, Mayne did not suggest that homosexuality had a corporeal basis to it, that homosexuals were hermaphroditic (though, as shown above, he did imagine rather hermaphroditic lesbians). Instead he used gradations in nature, including in sexual anatomy, to construe what he called "intersexuality"—what we now call homosexuality—as but one necessary element of "Nature's Endless Unity":

Between the whitest of men and the blackest negro stretches out a vast line of intermediary races as to their colours . . . Between a protozoan and the most perfect development of the mammalia, we trace a succession of dependent intersteps . . . A trilobite is at one end of Nature's workshop: a Spinoza, a Shakespeare, a Beethoven is at the other: led-to by cunning gradations, Nature can "evolve" an onion into a philosopher, or a mollusk to a prime-minister. The spectrum is a chain.[96]

"By what right," then, Mayne asked, have we

gone on insisting that each specimen of sex in humanity must conform absolutely to [one of] two theories [the "normal" male or "normal" female], must follow out [one of] two programmes only, or else be thought amiss, imperfect, and degenerate[?] Why have we set up masculinity and femininity as processes that have not perfectly logical and respectable inter-steps?[97]

Ulrichs and Mayne thus suggested a congenital origin of homosexuality and likened it to the anatomical hermaphroditism that arose occasionally in nature, in the hopes that the social status of homosexuality might be recalibrated and made acceptable because of its "naturalness."

Of course, such an analogy between the homosexual and the hermaphrodite left open the possibility for Mayne's contemporaries to draw two conclusions that Mayne did not wish to evoke: first, that "intersexuality" (here meaning homosexuality) was pathological, a "monstrous" state, like anatomical hermaphroditism; second, that the "intersex" (meaning homosexual), because it was a gradation between the two sexes, was at best an inferior form of life. Few of Mayne's contemporaries would have argued that the mollusk, though perhaps a necessary part of the Unity of Nature, was the prime minister's equal. The mollusk might have been natural, but by no means could it be accorded the status of a nation's leader. The hermaphrodite and the homosexual might be shown to be natural, but that did not mean they would be celebrated. Among most medical men, preventions and cures remained their preferred fate.

In France, where "sexual inversion was a terrifying nexus of medical, social and moral deviations,"[98] the "variation" of the invert ranked among inferior forms that needed explaining and treatment. Throughout the nineteenth century, French teratological theorists often used the notion of arrested development to explain a whole host of variations. ("Arrested development" was the term used to refer to the supposed failure of an organism to develop to its full potential, leaving it stuck in an "inferior" state.)[99] So it is not surprising that some French biomedical thinkers suggested that both anatomical hermaphroditism and "contrary" sexual desires originated from, or at least indicated, a "weakened procreative capacity," a failure of energy in the ascent up the sexual ladder.[100] Dr. Émile Laurent imagined in his 1894 text, *Les Bisexués* (The Bisexed), that a human being could slip down the sexual evolutionary scale in several ways: morphologically, in having confused sexual parts, as in the case of the traditional "hermaphrodite"; physiologically, in emitting the physiological signs of the "opposite" sex, as when a man lactated; or indeed, psychologically, in lacking the desire for the "opposite" sex or in evincing desire for the same sex. All these intertwined conditions represented "a return to inferior forms of hermaphroditism, a return to the monoecious creatures."[101] No mere natural varieties, these were devolutions, signs of "degeneracy" in both the physical and the moral sense.

Although Ellis concurred that "[s]trictly speaking, the invert is degenerate," he preferred not to use the term because of the regrettable pejorative nature it had acquired. According to Ellis, the invert was a "degenerate" only in the most scientific sense, namely that "he [*sic*] has fallen away from the genus."[102] He insisted that "in sexual inversion we have what may fairly be called a 'sport,' or variation, one of those organic aberrations which we see throughout living nature."[103] Ellis's opinions, like those of Krafft-Ebing, were clearly in the tradition of Ulrichs and Mayne, the tradition that held "that inverted sexual instinct was 'abnormal' but not a disease; it was instead a variation in a range of natural possibilities."[104] But there was always a serious problem for those who sought tolerance for homosexuality by likening it to hermaphroditism: hermaphroditism itself was considered thoroughly intolerable.

And, ultimately, so was homosexuality for the most part. Apparently that "homosexual" "hermaphrodite" Louise-Julia-Anna, whose story began this chapter, for several years after her encounter with Reumeaux and Guermonprez wandered northern France in search of a doctor who would allow her—help her—to go on loving men. Yet the doctors believed the ultimate truth and so her fate lay in her body, not in her desires, not in her acts. Louise-Julia-Anna was a man because she had testicles, and as a man she was a homosexual, and as a homosexual she had to be stopped.

Xavier Delore at the close of the nineteenth century summarized the situation thus: "From the scientific point of view, the *hermaphrodite* is very interesting; but from the practical point of view it has provoked . . . deceptive and equivocal situations."[105] Surely, he concluded, "the hermaphrodite is a trouble to society."[106]

CHAPTER 5

The Age of Gonads

In the nobler kinds, where strength could be afforded, her races are loyal to truth, as truth is the foundation of the social state.

—Ralph Waldo Emerson, "English Traits" (1856)

UNTIL NOW WE HAVE BEEN LOOKING mostly at those hermaphrodites labeled spurious, pseudo, or false. What happened to true hermaphrodites, beings who were by definition truly male and truly female, during this time? With the pervasive devotion among medical and scientific men to the one-body-one-sex rule, a devotion that seems to have grown more passionate with every empirical challenge, it should not come as much of a surprise that, by the end of the nineteenth century, true hermaphrodites for the most part ceased to be. I do not mean that the incidence of the particular anomaly—defined in the Age of Gonads as the coincidence of testicular and ovarian tissue in a single body—changed. Nor do I mean that true hermaphrodites were killed off, although we have seen that a few practitioners would not have been sorry to see hermaphroditism eliminated. Rather, true hermaphroditism started to disappear because medical and scientific authorities cooperatively revised the criteria for this condition such that it became almost impossible by the end of the century for any body to satisfy those guidelines.

I used to think that the study of the history of classification systems was as boring a job as one could take on. Even so, I knew that I had to trace the evolution of classification systems of hermaphrodites if I was to truly understand how medical and scientific men thought about hermaphroditism and if I was to get a sense of how scientific theory influenced medical practice, and practice theory. So I gathered, examined, and compared systems for classifying hermaphroditism—I collected the collections.

What I soon realized about these taxonomies was that they were incredibly interesting. They were, in an important sense, where half the action played. Figure out how someone organizes his world, and you will understand how he sees the world. You will also see how the organization system likely arranges the world in such a way as to reinforce that system maker's idea of the world—how what seems important gains in importance, how what seems unimportant fades from view. Londa Schiebinger has shown us, for instance, with a close analysis of the classification system of Carl Linnaeus, that we "mammals" got that name partly because of Linnaeus's fascination with women's breasts and his involvement in a debate over wet-nursing; Linnaeus wanted to see women stop farming out their babies to wet nurses, and this contributed to his decision to make the breast (the mammaries) the marker trait of mammals.[1] Mammals are thought to have many traits in common, but only the breast is used to construct our name. In the history of hermaphroditism, similarly, we see that late nineteenth-century medical and scientific men chose their particular marker—gonadal anatomy—for sex classification systems because it made it possible to uphold a stark division into female and male. It also greatly reduced the possibilities for true hermaphroditism in humans.

Two Early Systems

In contrast to the late nineteenth-century gonadal definition of true sex and true hermaphroditism, the two chief classification systems in use for hermaphroditism in France and Britain in the middle third of the century did not divide hermaphrodites strictly along gonadal lines. The architects of those two earlier systems were Isidore Geoffroy Saint-Hilaire and Sir James Young Simpson. Before the adoption of the gonadal criteria for the classification of hermaphroditism, most French medical and scientific men used Geoffroy's system, and most British medical and scientific men used Simpson's.

We have already met Isidore Geoffroy Saint-Hilaire (1805–1861), that ambitious teratologist and faithful son of the renowned anatomist Étienne Geoffroy Saint-Hilaire. Recall that all of Isidore Geoffroy's teratological classification systems reflected his belief that "nature is one whole," that is, that all variations, no matter how "monstrous," were explainable through reference to normal development. Most if not all monstrosities, in Geoffroy's eyes, were rooted in excesses and arrests of otherwise normal development. (Thus, for instance, extra fingers represented excesses of normal development.) With this, his so-called anatomical philosophy, as an unshakable tenet, Geoffroy organized variations along scales ranging from the relatively minor to the most dramatic, for he believed that doing so would reveal the steps of the developmental ladder and would elucidate the origin of both normal and abnormal anatomy.

Geoffroy thought hermaphroditic variations so unusual and interesting that he dedicated one of his four great teratological kingdoms to hermaphroditisms alone.[2] He first presented his analytic classification system for hermaphroditism at the Académie des Sciences in Paris in 1833.[3] This was literally an analytic system—that is, Geoffroy suggested that the sex of a body should be analyzed in terms of the body's sexual organs and the normal or abnormal relations of those organs. Normal malehood, normal femalehood, and various kinds of hermaphroditism were each to be equated with a particular combination of organs.

To make his system as rigorous as possible—he sought a quantitative certainty in teratology—Geoffroy decided to divide human sexual anatomy into six "sex segments." The six segments included the left and right segments of each of three anatomical sex zones, which he termed: (1) the "profound portion," which included the ovaries or testicles and their anatomical "dependents"; (2) the "middle portion," which included the uterus or prostate and the seminal vesicles, and their anatomical "dependents"; and (3) the "external portion," which included the clitoris and vulva or penis and scrotum. According to Geoffroy's system of sex classification, one could determine the malehood, femalehood, or kind of hermaphroditism of a given subject by analyzing the sexual organs found in each of that subject's six sex segments. Were any of the segments male? Were any female? Were any "hermaphroditic," that is, inclusive of mixed or ambiguous organs? In Geoffroy's system, if all of a subject's six segments included the normal male parts, then that subject was a male. If all of the six segments included the normal female parts, then that subject was a female. A juxtaposition of male and female segments or the presence of any ambiguous parts constituted a hermaphroditism.

At the general level Geoffroy classified hermaphrodites as follows:[4]

I. First class, without excess in the number of body parts:

Order 1: Masculine hermaphrodism: sexual apparatus essentially male.

Order 2: Feminine hermaphrodism: sexual apparatus essentially female.

Order 3: Neuter hermaphrodism: sexual apparatus presenting some intermediary conditions between those of the male and those of the female, and being really of no sex.

Order 4: Mixed hermaphrodism: sexual apparatus in part male and in part female.[5]

II. Second class, with excess in the number of body parts:

Order 1: Complex masculine hermaphrodism: [complete] male sexual apparatus with some supernumerary female parts.

Order 2: Complex feminine hermaphrodism: [complete] female sexual apparatus with some supernumerary male parts.

Order 3: Bisexual hermaphrodism: A male and a female sexual apparatus.

(a) Imperfect bisexual hermaphrodism: one or both of the apparatuses incomplete.

(b) Perfect bisexual hermaphrodism: two complete apparatuses.

Note that, significantly, Geoffroy did not make his primary division of the hermaphrodites into "true" and "pseudo," the primary division that would pervade the last third of the nineteenth century. Instead Geoffroy divided hermaphrodites primarily into those without excess in the number of body parts and those with excess in the number of body parts.[6] And, among his fourth order of the hermaphroditisms without excess in the number of body parts, Geoffroy explicitly included several types in which there were found both ovaries and testes in a single individual. This seemed to Geoffroy not only a conceivable possibility, but one of which there had been recorded cases in humans and other higher vertebrates.[7] Like many of his successors, Geoffroy posited that it was the "interior parts . . . which essentially characterized the sex."[8] Geoffroy's system of hermaphroditic variations did not rest, however, upon a rigorous notion of, or obsession with, a "true" gonadal sex of each hermaphroditic body, and Geoffroy did not share the latter-day obsession with "true" versus "false" hermaphrodites.

The man who gave British medicine its long-standing classification system for hermaphroditism was not by trade a museum-based scientist like Geoffroy, but was rather a practicing obstetrician. When James

Young Simpson, M.D. (1811–1870), provided his lengthy article "Hermaphroditism, or Hermaphrodism"[9] to Robert Bentley Todd's *Cyclopaedia of Anatomy and Physiology* in the late 1830s, he was a Fellow of the Royal College of Physicians and lecturer on midwifery at Edinburgh. Simpson's classification system was far less rigorous and philosophical than Geoffroy's. Simpson opted simply to group types of hermaphroditism into readily comprehended categories. No translating from an abstract type was required as it was in Geoffroy's system. Also in contrast to Geoffroy, Simpson divided hermaphroditism into the two orders of

> *Spurious* and *True;* the spurious comprehending such malformations of the genital organs of one sex as make these organs *approximate in appearance* and form to those of the opposite sexual type; and the order, again, of true hermaphroditism including under it all cases in which there is an *actual mixture or blending together,* upon the same individual, of *more or fewer* of both the male and female organs.[10]

Note that the criteria for "true" hermaphroditism, as laid out by Simpson and generally adopted thereafter in Britain, were fairly broad. Beings with many different combinations of "male" and "female" parts counted as true hermaphrodites.

More specifically, Simpson classified the types of hermaphroditism as follows:[11]

I. Spurious hermaphroditism: . . . the genital organs and general sexual configuration of one sex approach, from imperfect or abnormal development, to those of the opposite . . .
    A. In the female:
        1. From excessive development of the clitoris, etc.
        2. From prolapsus of the uterus.
    B. In the male:
        1. From extroversion of the urinary bladder.
        2. From adhesion of the penis to the scrotum.
        3. From hypospadic fissure of the urethra, etc.
II. True hermaphroditism: . . . there actually coexist upon the body of the same individual more or fewer of the genital organs and distinctive sexual characters both of the male and female . . .
    A. Lateral true hermaphroditism:
        1. Testis on the right, and ovary on the left side.
        2. Testis on the left, and ovary on the right side.

B. Transverse true hermaphroditism:
    1. External sexual organs female, internal male.
    2. External sexual organs male, internal female .
C. Vertical, double, or complex true hermaphroditism:
    1. Ovaries and an imperfect uterus with male vesiculae seminales, and rudiments of vasa deferentia.
    2. Testicles, vasa deferentia, and vesiculae seminales, with an imperfect female uterus and its appendages.
    3. Ovaries and testicles coexisting on one or both sides, &c.

Note that some hermaphrodites classified as "true" under Simpson's system had what appeared to their investigators to be both ovaries and testes, although, unlike later in the century, independent and microscopic verification of the gonads' nature was generally not required. But additionally, and of particular importance for our purposes, under Simpson's classification scheme, true hermaphrodites were not required to possess both types of gonads. For example, a person provided with testes and a uterus but no ovaries could qualify as a true hermaphrodite.[12]

Indeed Simpson and his contemporaries counted as a true hermaphrodite the Italian Maria Arsano, mentioned in Chapter 2, although Arsano was not said to have any female organs internally (see Figure 10). Arsano was the posthumously famed hermaphrodite who lived for eighty years as a woman, only to have her "male" internal anatomy discovered by chance after death. Observers described Arsano as externally female and internally male, and, in the 1830s until about the 1870s, this combination of a "feminine" exterior and a "masculine" interior was enough to make her "truly" hermaphroditic. In other words, Arsano was, before the Age of Gonads, a "true hermaphrodite" although—or rather because—internally she was thoroughly male.

There was no notion in Simpson's system of a "true sex" which was signaled exclusively by the gonad, nor was there the notion that the true hermaphrodite was only that very rare being who had both testicles and ovaries. Even Simpson's contemporary, the well-known surgeon Robert Knox, who criticized Simpson's work because of philosophical disagreements, nonetheless concurred that the useful definition of true hermaphroditism was simply "the existence of [various] male and female organs in the same individual . . . these organs [being] more or less perfect."[13] Many beings therefore counted as true hermaphrodites before the Age of Gonads.

The Rise of the Gonads

The classification system for hermaphroditisms presented in 1876 in the *Handbuch der Pathologischen Anatomie* by Theodor Albrecht Edwin Klebs (1834–1913) functioned something like a constitution for the Age of Gonads. Klebs's taxonomy, apparently the first of its kind, codified the belief that true sex should be based exclusively on the nature of the gonads. Klebs studied medicine in Berlin, and although he was originally interested chiefly in obstetrical practice, under the guidance of Rudolf Virchow he instead turned his interests to pathology.[14] In his 1876 text on pathological anatomy, Klebs divided hermaphroditism into two basic kinds, true and false, based on the gonads, and then he subdivided those kinds:[15]

I. True hermaphroditism (presence of the ovaries and testes in one individual):
    (a) true bilateral hermaphrodism: one testicle and one ovary on each side of the body.
    (b) true unilateral hermaphrodism: on one side an ovary *or* a testicle, on the other an ovary *and* a testicle.
    (c) true lateral hermaphrodism: (also called alternated hermaphrodism) an ovary on one side, a testicle on the other.
II. Pseudohermaphroditism (spurious hermaphrodism; "doubling of the external genital apparatus with a single kind of sexual gland"):
    (a) masculine pseudohermaphroditism: presence of testicles and evident development of the feminine genital parts. [16]
    (b) feminine pseudohermaphroditism: presence of ovaries with some predominance of the masculine genital parts. [17]

Note that one of the significant results of Klebs's system was that a being could appear almost entirely feminine internally and externally and still be considered a true male by virtue of the possession of testicles and a lack of ovaries. Similarly, a being could look and act very masculine but would have to be classified as female if he had ovaries. So, for instance, under Klebs's system, Maria Arsano, previously a true hermaphrodite, was reclassified as a masculine *pseudo*hermaphrodite. Similarly, L.S., the Parisian model introduced in Chapter 4, would have been a true hermaphrodite under Simpson's system but became a masculine pseudohermaphrodite under Klebs's system.

In short, under Klebs's system, significantly fewer people counted as "truly" both male and female. This is the trend that we see throughout the rest of the period, that is, the trend toward the elimination of true hermaphroditism in humans. Until the mid-1880s in Britain, many established practitioners continued to employ fairly liberal definitions of true hermaphroditism. In 1885, for instance, the Glasgow surgeon Buchanan considered little Christina a true hermaphrodite although he did not think she had any ovaries to match her testicles. And in that same year, Fancourt Barnes and his father Robert provided in their obstetrical textbook a fairly vague, Simpsonesque description of what determined a case of true hermaphroditism, saying it was one "in which some of the sexual organs of both sexes existed in one individual."[18] But increasingly through the 1870s, 1880s, and especially the tumultuous, hermaphrodite-ridden 1890s, more and more medical men grew impatient with definitions of true hermaphroditism that paid mind to any criterion other than the establishment of both ovarian and testicular tissue.[19]

In both France and Britain, medical and scientific men moved steadily toward the explicit use of the gonads as the sole markers of true sex. Interestingly, Klebs's classification system apparently was not the catalyst for the shift in Britain and France toward the gonadal definition of true sex in the 1870s and 1880s, because neither Klebs's ideas nor his classification system appears to have been cited until the 1890s in any of the British and French literature. Instead it appears that the ultimate adoption of Klebs's classification system in France and Britain in the 1890s occurred simply because Klebs's system codified a by then already widely accepted notion, that is, that the gonads alone mark true sex.

In the 1890s, Klebs's system came to be known in France and Britain via the publications of Samuel Pozzi and Franz Neugebauer and via a watershed article published in the 1896 *Transactions of the Obstetrical Society of London*.[20] That critical article was coauthored by two men, George F. Blacker (1865–1948), assistant obstetric physician to University College Hospital, and Thomas William Pelham Lawrence (1858–1936), curator of the museum at University College London. In their polemic, Blacker and Lawrence set out to tighten the definition of true hermaphroditism and to clean the historical record of any alleged cases of true hermaphroditism that did not fit their refined, stricter definition. Specifically Blacker and Lawrence insisted that, regarding supposed cases of true hermaphroditism, the utmost importance is to be laid upon the presence or absence of the genital glands themselves; and to establish their presence we must insist upon the necessity of a *microscopical examination.* Without this it is quite impossible to say what the exact nature of any gland may be.[21]

While this new criteria did not eliminate the possibility of true hermaphroditism in humans—in fact Blacker and Lawrence were "convinced . . . that true hermaphroditism can occur and has occurred in man" and they themselves offered an original case—the tougher prerequisites did radically reduce the number of true human hermaphrodites.[22]

Blacker and Lawrence actually had to adapt Klebs's system of classification to fit their own case. The definition of true hermaphroditism suggested by Klebs assumed that in a true hermaphrodite the ovaries and testicles would be separate organs. When Klebs had put forth his definitions, there were no recorded cases of ovotestes. But Blacker and Lawrence discovered in their own subject, an eight-and-a-half-month stillborn fetus, what they thought to be an "ovotestis," that is, an organ which contained both ovarian and testicular tissue. Blacker and Lawrence confirmed this with microscopic analysis and decided that their case fit the category of "true unilateral hermaphroditism" described by Klebs, because their subject had ovarian tissue on one side of the body, and ovarian and testicular tissue contained in a single organ on the other side.[23] (See Figure 20.)

Blacker and Lawrence took several pages to describe their interesting new case, but the bulk of their article was dedicated to a close scrutiny of all prior alleged cases of true hermaphroditism—"true" as they understood Klebs to have defined it, that is, having present both ovarian and testicular tissue. Blacker and Lawrence did not bother to review any case in which both kinds of tissues were not alleged to be present. (Maria Arsano's case, for example, would not have even been considered, because she had only testes.) Of the twenty-eight cases Blacker and Lawrence did review, they eliminated from the ranks of true hermaphroditism all but three. Alleged cases were disallowed: if no microscopic examination had been made;[24] if the microscopic examination was poorly performed or if it was contradicted by an original drawing or wax replica, or by a new examination of the preserved specimen; if the reporting was scanty, shoddy, or unbelievable; if Blacker and Lawrence could not find the original description of the case.[25] Now wholly convinced that one

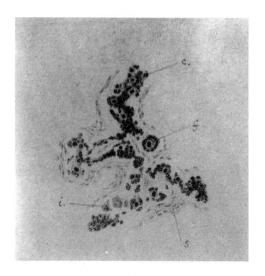

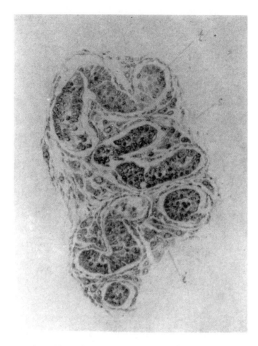

Figure 20 Photographs of histologic (microscopic) sections of a gonad, provided by Blacker and Lawrence as evidence of their alleged case of true hermaphroditism. Blacker and Lawrence claimed their subject's left sex gland (shown here in two sections) was an "ovotestis." The top section shows the "ovarian portion," and the bottom shows the "testicular portion." Blacker and Lawrence did not provide an illustration of any gross anatomy.

could not declare true hermaphroditism without strict microscopic analysis of the gonads and clear evidence of testicular and ovarian tissue, Blacker and Lawrence expressed amazement to discover "upon what extremely slight foundations the diagnosis of true hermaphroditism was at one time based."[26]

The new evidentiary criteria cemented by Blacker and Lawrence's very influential treatise had three major outcomes in both France and Britain. First, medical history was thoroughly revised, as nearly all previous cases were struck from the already dwindling ranks of so-called true hermaphrodites for failure to meet or evidence the new criteria. After 1896, nearly every documenter of hermaphroditism adhered to the new gonadal and histological criteria, and it now became in the fashion of Blacker and Lawrence a popular endeavor of those writing of sexually ambiguous subjects to review cases in the medical record.[27] Authors had done this in the earlier part of the century too, but then it was to compare their own cases to prior ones like them for diagnostic and illustrative (and sometimes entertainment) purposes. Now investigators reviewed past accounts in order to reclassify the conditions, to move great numbers of previously true hermaphrodites into the category of the spurious.[28] (Ironically, even Blacker and Lawrence's own case fell victim within a few years to this reassessment blitz when successors suggested Blacker and Lawrence had misinterpreted their specimen's histological characteristics.)[29]

The second result of Blacker and Lawrence's rules was that far fewer new cases met the criteria for true hermaphroditism, and so now almost all newly discovered hermaphrodites were considered "spurious," no matter how confused their traits (or their doctors). Now a practitioner could also, as we have seen, by a single criterion decide who was "truly" a man and who a woman. This very strict conceptual order was imposed on any being who seemed otherwise likely to threaten the borders between males and females.

The third outcome of Blacker and Lawrence's new standardized rules was that, given the limitations of medicine at the time—that is, given that biopsies were not performed on hermaphrodites in France and Britain until the 1910s—the only possible true hermaphrodite was a dead and dissected hermaphrodite or, at best, a castrated one.[30] After all, the only way one would convincingly provide evidence of true hermaphroditism was by removing, slicing, and microscopically examining the whole gonad. Finally, as if to take away any power remaining in the bodies of true hermaphrodites, medical men now also speculated that, even when a truly sexually mixed being did exist, rather than being doubly potent, it was sure to be thoroughly impotent. Jonathan Hutchinson declared in 1896 his conviction that "whenever [true hermaphroditism] is the case, the organs are never developed in perfection. The testes do not secrete semen, or the ovaries do not attain their functional activity, and thus the being who is in a sense bisexed is in another sense unsexed, and never attains the full development of either."[31] Across the Channel, in the *Annales d'hygiene publique et de médecine légale* (Annals of Public Hygiene and Medical Jurisprudence), Pierre Garnier painted a similarly impotent image of the true hermaphrodite human:

> All the cases [of human hermaphrodism] do not evidence true hermaphrodism [in humans] as it is encountered . . . in the brilliant and perfumed world of the flowers. [True hermaphrodism] scarcely exists in the human species, a type with distinct sexes separated at the physical and at the moral [levels]. The person uniting partially the organs of the two sexes ordinarily does not enjoy any. Instead of being at once man and woman, it is neither one nor the other; it is neuter, or rather a monster.[32]

Small in number, dead, impotent—what a sorry lot the true hermaphrodites had become! And what a strange and remote look true hermaphroditism now had. No longer was the condition typically pictured in bodies, faces, or even genitalia. Now, in the 1890s, true hermaphroditism was generally depicted as in Figure 20—as tissue samples. By the end of the nineteenth century, the gonadal definition of true sex meant that "truth" was determined by the nature of the gonads, even if that "truth" were invisible and unsuspected in a living patient. Furthermore, the only true hermaphroditism would exist on a microscope slide after the death or castration of the person from whom the sample came.

Motivations for the Gonadal Definition

What was the logic behind the claim that true sex—and therefore true hermaphroditism—should be determined solely by the anatomical nature of the gonads? Given Klebs's training as a pathologist, perhaps it should be no surprise that he would opt for a histological (tissue-based) approach to hermaphroditism. Blacker, Lawrence, and their colleagues were clearly influenced by German pathologists like Klebs and by the international growing interest in laboratory-based medical research.[33] But why such an intense, communal focus on the gonads specifically as the sole markers of true sex? We might guess that, like Klebs, French and British theoreticians including Blacker and Lawrence had in mind some subtle reasoning related to the embryological differentiation of the sex organs in males and females.[34] As discussed in Chapter 2, the gonads were known to emanate early in development from the genital ridge, and to differentiate relatively early into proto-ovaries and proto-testes. Among embryologists of the time, the earlier the development of a given structure, the more important that structure was understood to be. So of all the many sex traits supposedly distinctive to men and women, it might seem that the testes and ovaries were the ones which, even very early in development, could mark a person's ultimate "true" sex. This possibility—that the gonads were chosen as the ultimate markers of true sex because they emerged early in embryological development—is an interesting one, but oddly, French and British texts written on hermaphroditism during this period make no such embryological arguments for the gonadal definition. If this was the logic, it was never stated.

I think it more likely that the choice of the gonads as markers of true sex derived from the late nineteenth-century idea that the fundamental difference between men and women lay in their reproductive capabilities. Typical for the time, in his 1892 textbook of human embryology, Charles Sedgwick Minot, then professor of histology and human embryology at the Harvard Medical School, declared with reference to human sex: "The structure and functions of the genital ducts, of the uterus, mammary glands, etc., though eminently characteristic of the sexes, in man [sic] are *not* from a biological point of view fundamental."[35] Minot argued that instead the "essential difference" between the male and the female is "the production of the genoblasts; the male produces the spermatozoa, the female the ova."[36] Similarly, in 1888 Dr. Jean Tapie, professor at the School of Medicine in Toulouse, insisted overtly in his study of hermaphroditism that "the truly characteristic organs of sexuality are those which furnish the elements necessary to reproduction, the testicle in the man, the ovary in the woman."[37] In late nineteenth-century biomedical texts on sex differences, reproductive difference was most often cited as *the* fundamental difference between men and women, the very thing that defined Man and Woman, the thing that was in turn the source of all their other differences.[38]

Yet, as we have seen repeatedly, despite Tapie's and his colleagues' apparent reproductive-role logic for the gonadal definition, medical men were in fact unconcerned with the reproductive capabilities of any particular subject when it came to "true sex" categorization. If different reproductive capabilities were what defined Man and Woman, they were not what defined, in doubtful cases, "true men" and "true women" or "true hermaphrodites." Even subjects anatomically incapable of impregnating a woman were considered true males because they had testicular tissue, and even subjects anatomically incapable of being impregnated were considered true females because they had ovarian tissue. In actual cases of hermaphroditism, the reproductive capabilities of doubtful subjects was thus not at issue in terms of diagnoses of true sex. Anatomy of the gonads was all that mattered.[39]

Perhaps then it was the physiological role of the gonads with respect to secondary sexual characteristics that made them seem so all-important to a person's "true" sex in the minds of medical and scientific men. The subject period did see the rise of internal secretion theories and the study of ductless glands (what would several decades later become endocrinology). Charles Edouard Brown-Séquard (famous for injecting himself with testicular derivatives) and many other scientists throughout Europe widely publicized their findings with regard to the systemic effects of testicular and ovarian extracts, and dozens of investigators had observed the changes in sexual characters that took place after the removal of the gonads in people and animals.[40] The gonads were well on their way to being understood as sex *glands*. But the first overt reference to internal secretions (later isolated and called hormones) as the logic for the strict true-hermaphroditism criteria came only in 1906, a full decade after the general acceptance of the new criteria and nearly three decades after its initial codification. Indeed, in 1896 Blacker and Lawrence had made it absolutely clear that they were unconcerned with functionality, writing that true hermaphroditism was to be "taken to mean merely the presence of different genital glands in one individual, without reference to the presence or absence of their functional activity."[41] Moreover, hermaphroditism seemed to defy the very core of internal secretion theory, because some spurious hermaphrodites with testicles displayed entirely female secondary-sexual characteristics, and vice versa. So attempting to base the criteria on physiological aspects of the glands at this time would only have confused the issue.

What is so striking is the very strict yet somewhat arbitrary character of the new, reductionistic definition of sex in the context of true hermaphroditism. Medical and scientific men might well have stuck to a broader (or for that matter, stricter) definition of true hermaphroditism and true sex.[42] I therefore suspect that the widespread adoption of the gonadal definition of sex was driven not by a strictly "scientific" rationale but instead for the most part by pragmatism: it accomplished the desired preservation of clear distinctions between males and females in theory and practice in the face of creeping sexual doubt. The practical result of the adoption of the gonadal definition was that most bodies, no matter how ambiguous looking or acting, were entitled only to a single sex, and "true" living hermaphrodites were—by definition—impossible.

In trying to understand the revision of true hermaphroditism as a concept, it certainly would be a mistake to neglect the inevitable importance of the concurrent rise of social challenges to sex borders. Let me again suggest that, for instance, it cannot be a coincidence that at the same time other historians find the emergence of the homosexual, I find the virtual extinction of the hermaphrodite. And surely the hermaphroditism experts must have known of feminism as something more than the anatomical pathology for which they used the term ("the arrest of development of the male towards the age of puberty, which gives to it somewhat the attributes of the female").[43] For a medical man to admit a living, doubtful subject to true hermaphroditism would have been potentially to add to the threat of social sex confusion fomented by people like feminists and homosexuals.

Perhaps this helps explain the theoretical single-mindedness by the 1890s in matters of true hermaphroditism among men who, in medical practice and discourse, frequently disagreed about the nature and limits of males and females. We must remember that the business of distinguishing and keeping clearly separated "true" men, "true" women, and "true" hermaphrodites was never seen as a merely academic or as an isolated exercise. Definitions of "true" and "spurious" hermaphroditism and "true sex" always carried with them political implications. Even Blacker and Lawrence, though apparently primarily concerned with strictly histological matters, nonetheless recounted the scandalous sexual exploits of various previously "true" hermaphrodites and bemoaned the fact that

at one time the term [true hermaphroditism] was used with great laxity to include all cases where the genitalia of an individual in any respect resembled those of the opposite sex. As a result of this, many persons were termed hermaphrodites without just cause; and while women with beards were frequently called males, men with well-developed mammae were relegated to the class of females.[44]

Blacker and Lawrence believed that medical men had to do a better job of keeping true men and true women correctly labeled. Too often, it seemed, an "error of nomenclature" had occurred, leaving a true man in the role of a woman or a true woman in the role of a man. It seemed as if the crystal-clear gonadal definition of sex would restore order in the laboratory, in the surgical clinic, in marital beds, in military barracks, on the streets.

Indeed, the more confusing and abundant cases of doubtful sex became, the more certain and constrictive concepts of true hermaphroditism grew. If practitioners could barely keep their heads above the doubtful sex that walked into their hospitals and offices, at least conceptually sex in the realm of true hermaphroditism seemed safely confined and certain. In the late 1890s and early twentieth century, the confident, reverberating declarations by medical men about the nature of true hermaphroditism, and by implication the nature of sex, amid all the blooming sexual confusion they witnessed, was like the frantic painting of a house falling down. Medical men triumphantly reconstructed and constricted the true hermaphrodite even while—perhaps because—they witnessed and were forced to confess serious doubts about sex.

Further Attempts to Eliminate "True Hermaphroditism"

It was perhaps inevitable that the discomfort with the idea of human true hermaphroditism would reach the point where medical men would begin to argue that the very word "hermaphrodite" needed to be done away with altogether. Advocates of this linguistic cleansing appeared all over Europe, but I consider here the arguments of two, Samuel Pozzi and David Berry Hart.

By the time Samuel Pozzi offered his refined classification system and nomenclature for what had been called hermaphroditism to the Académie de Médecine in 1911, a clear pattern had been documented regarding pseudohermaphrodites' ovaries and testes: hermaphrodites' gonads were often abnormally developed such that they did not appear to be fully functional in terms of chemical secretions.[45] Now Pozzi, who had always held the conviction that there could be no true hermaphroditism among humans, appealed to these discoveries regarding the physiologic functionality of hermaphrodites' gonads to announce in 1911 that even so-called true hermaphrodites with ovotestes were not really true hermaphrodites because, Pozzi believed, only the ovarian portion of their ovotestes functioned.[46] "Thus," Pozzi declared, "there are no true hermaphrodites, but there exists a rather frequent anomaly which can give the *illusion* of bisexuality. Individuals who present it actually merit the name of pseudohermaphrodites."[47]

Considering that Pozzi eliminated the category of true hermaphrodites using a criterion of sex gland functionality, one might expect that he would also consider sex gland functionality when categorizing pseudohermaphrodites. But he did not. Instead, Pozzi and his colleagues continued to cling to the strictly anatomical criterion of gonadal tissue to distinguish "true" males from "true" females in cases where there were only testes or only ovaries.[48]

The consequence of this line of thinking was paradoxical. True hermaphrodites were now to be considered not really true, because they did not have both functioning ovarian tissue and functioning testicular tissue. Yet the diagnosis of "true sex" (as female or male) in pseudohermaphrodites did not at all depend on functionality of the glands. This meant that a single sex could be easily assigned to every ambiguous pseudohermaphrodite, and that every true hermaphrodite would now be stripped of its claim to being "truly" both male and female. In short, the contradictory conceptual treatment applied to "true" hermaphrodites and "pseudo" hermaphrodites had one common goal: to eliminate the idea once and for all that a human could truly possess two sexes. Under Pozzi's new system, no matter how bizarre a body looked or acted, it still was assigned only one sex because the bilateral criteria for true hermaphrodites and for pseudohermaphrodites would allow for one and only one sex per body.

In his 1911 discussion of hermaphroditism, Pozzi noted that most medical men faced with cases of pseudohermaphroditism relied on the classification system of Klebs.[49] But Pozzi found Klebs's system faulty and clumsy: faulty in that it admitted true hermaphroditism, and clumsy in that, at least when the system for pseudohermaphroditism was translated into French, the objects to which Klebs's adjectives "internal" and "external" referred became unclear.[50] Pozzi preferred and promoted his own system. Since he had conceptually eliminated true hermaphroditism, he did not bother to pay it any mind in his classification scheme. Instead, the system he proposed split pseudohermaphrodites along—and thereby re-inscribed—the great divide: in one great class stood the true males, in the other, the true females.[51] Shying away from the troubling term "hermaphroditism," Pozzi suggested a new phraseology for people with mixed sex traits. True men with womanly characteristics were to be labeled "androgynoids," and true women with manly characteristics, "gyn-androids." The first half of the word indicated the "true" (gonadal) sex; the second half signaled the apparent ambiguity.[52] (Not surprisingly, these awkward terms did not catch on.)

Three years later, in 1914, equally frustrated with the term "hermaphroditism," David Berry Hart (1851–1920) also called for the elimination of the word from the medical lexicon. Hart, then Fellow of the Royal College of Physicians and lecturer on midwifery and the diseases of women in Edinburgh, disbelieved all alleged accounts in which testicular and ovarian tissue were both found in a single human and so concluded that even the term "pseudohermaphroditism" should be abandoned, because "we cannot have a pseudo form of a non-existent condition."[53] Hart was a man fascinated by sex difference,[54] and in his work, the isolation of "true," that is, gonadal, sex from all other "sexual" characteristics became absolute—as did, by consequence, the profound distinction between men and women.

Nonetheless, Hart proposed the term "sex-ensemble" to designate all those anatomical characteristics typically associated with sex. The sex gland would remain "the only criterion of sex," and any subject's true sex therefore quite a simple matter, as simple and basic as the supposed differences between men and women.[55] Meanwhile, those conditions formerly known as "pseudohermaphroditism" would simply constitute "atypical presentations of the sex-ensemble." According to Hart, the "typical female sex-ensemble" was

> made up of (a) the ovary; (b) the potent sex-duct tract—[fallopian] tubes, uterus, vagina, and external genitals; (c) the [rudimentary] opposite sex-duct elements—epoöphoron, degenerated equivalent of the epididymis of the male; (d) the secondary and congruent sexual characters—hair distribution, pelvis, body form, vocal cords, ossification of thyroid cartilages (incomplete), and the psycho-sexual feeling for the male; mentality less strong than in male.[56]

By contrast the "typical male sex-ensemble" comprised

(a) descended testes; (b) vas deferens and phallus—the potent organs; (c) the [rudimentary] opposite sex-duct elements—hydatid testis and prostatic utricle; (d) the secondary sexual characters—hair distribution, pelvis, body form, vocal cords, ossification of thyroid cartilages (complete); and the psycho-sexual feeling for the female; mentality stronger than in female.[57]

Hart noted that atypical sex-ensembles often arose because of an abnormality in the sex glands.[58] Yet no matter how dysfunctional a sex gland might be, no matter how much the sex-ensemble might recall the sex "opposite" to the gonadal sex, the true sex of the individual remained gonadally—purely anatomically—determined. Sex gland tissue—functioning or not—would still signal and enforce the distinctions between men and women. And "hermaphrodites" would be gotten rid of altogether, fit into the realm of the two sexes, even if they were "atypical" examples of their true sexes.

The Shift to Effective Sex and to "Gender"

But should everyone—no matter what one looked like, acted like, or felt like—be socially designated the sex of one's gonads? Indeed, is this what occurred?

Not always. As we have seen, particularly in the last chapter, British and French medical men sometimes let a case of "mistaken sex" go, and sometimes even worked to create a sex that matched the social sex rather than the gonadal sex of a doubtful patient. Recall the British cases of Christina, A.H., and the Middlesex widow, all of whom had their testicles removed. Perhaps in these surgeons' minds the removal of incongruous organs served to reinforce the sex that seemed the true one, even if true sex theoretically depended specifically on the nature of the gonads. With the one-body-one-sex rule, perhaps it seemed easier and better to take out surprising organs than to claim that a woman was a man, or a man a woman.

Although French surgeons and physicians apparently did not participate in such "desexing" (or resexing) procedures, they did sometimes let a case of mistaken sex go or offered to try to alter genitals to better match the social sex.[59] Clearly, while in theory medical men may have all agreed that sex was determined by the nature of the gonads, in practice—particularly in Britain—outward appearances mattered very much. Indeed, every encounter between a hermaphrodite and a medical man demonstrates this point, for whatever the degree of attention focused on the gonads, much heed was also paid to other "sex" traits—not only as indicators but also occasionally as determinants of ultimate sex. Thus, for instance, in 1888 when a Birmingham surgeon was presented with two-year-old Sarah Jane and her sizable penis or clitoris, he could not be sure if the child was gonadally male or female, "but after carefully considering the case in all its bearings, *social as well as anatomical,* [the surgeon] accepted the mother's suggestion, and amputated the member close to the pubes, and the child made a rapid recovery, and is being brought up as a female."[60] Sarah Jane may still have had testicles, but she was allowed to leave a girl.

So, as was true for the status of so many other supposedly "sexed" traits, despite medical theory and rhetoric, the status of the gonads was not, in practice, completely clear or fixed. It seems that, when revealed, the sex glands counted as sufficient signs of sex in most doubtful cases, and yet they were not always considered necessary signs. If gonads had been the only trait that really counted in determining sex, then the sex of all hermaphrodites would have been automatically changed to equal the sex of the gonads, and men with breasts, girls with large erectile organs, people with the "wrong" sizes, shapes, sorts of hair, chests, voices, and desires would not have posed the trouble they did. While gonads held important agency (diagnostically and physiologically) in the realm of sex, throughout this period many other traits retained deep—if enormously contested—meaning. Medical and scientific men cheered on the gonadal definition of sex which required all true hermaphrodites to have microscopically verified testicular and ovarian tissue, but the application of the gonadal dictum in practice was at best uneven.

Yet, curiously, until 1915 when William Blair Bell did so, apparently no medical man dared to openly question the gonadal definition of true sex. Blair Bell seems to have been the first among the British and French medical establishments to overtly—if tentatively—suggest that in actual cases of unusually mixed sexual characteristics, the "true sex" ought not be judged by the often (apparently) dysfunctional sex gland—that such a practice made little sense. A good part of what pushed Blair Bell to make this suggestion was the rise of new medical technologies: laparotomies (exploratory surgeries) and biopsies (removal of small samples of suspicious tissues) by 1915 made it possible in practice to absolutely confirm testes in living women, ovaries in living men, and ovotestes in living true hermaphrodites. Blair Bell therefore suggested that the time for a gonadal definition of every body's true sex had now passed.

Blair Bell (1871–1936), staff surgeon to the Royal Infirmary at Liverpool, made his recommendations before a meeting of the Liverpool Medical Institution on February 11, 1915. There he recalled that "since the scientific investigation of these cases [of hermaphroditism] began, the custom has been to define the sex of the individual according to the sex-characteristics of the gonads."[61] But two particular recent cases gave him pause in this regard, and Blair Bell wondered aloud whether some revision in the practice of sex diagnosis was not badly wanted. One was a case in which a patient had an ovotestis; the second was a case of what today would be called androgen insensitivity syndrome, that is, the case of a very "womanly" patient who was shown to have two testes and no ovaries. According to the strict gonadal definition of true sex, the former would be a confirmed living true hermaphrodite, the latter an astonishingly womanly man.

The first case to trouble Blair Bell was that of S.B., who was seventeen when first examined in November 1912. By that time, S.B.'s menses, which had started three years earlier, had been absent for eighteen months, and her voice had dropped.[62] Her thyroid appeared enlarged. "A diagnosis of suprarenal hyperplasia was made, and the patient was treated with ovarian and thyroid extracts for some time."[63] When seen again in August of 1914, she appeared still more masculinized (see Figure 21). "She had a slight moustache and a masculine distribution of the hair on the body. There was still complete amenorrhoea." Additionally, her clitoris had grown to two inches, and was found to be covered with "a well-marked prepuce." The left genital gland felt enlarged, but the suprarenal region seemed normal. S.B. had now failed to menstruate for three years.[64]

Suspecting an ovarian tumor, Blair Bell performed exploratory surgery in September 1914, and took tissue samples from both "ovaries." The left seemed at first examination to be abnormal. Before closing, Blair Bell implanted in S.B.'s uterus "a small graft from an ovary removed from a patient operated upon a few minutes before," with the aim of providing functional tissue to S.B., whose ovaries presumably were failing to operate normally.[65] The "ovarian" sections taken from S.B. were then sent to the pathological laboratory of the hospital, and the report came back that the left ovary showed "what is undoubtedly a columnar-celled carcinoma

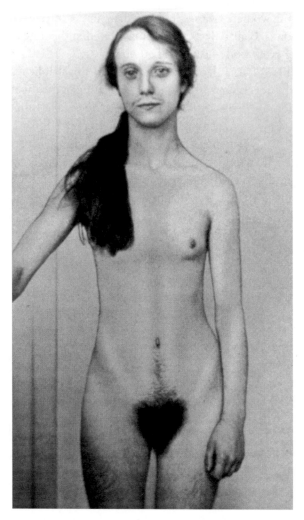

Figure 21 Photograph of S.B. taken in August 1914, at nineteen years of age.

[cancer] with well-marked acini."[66] Blair Bell reported in his presentation to the Liverpool Medical Institution:

> Placing an unwise faith in this report, I reopened the abdomen on the 22nd September 1914, and removed both ovaries, tubes, and the fundus of the uterus. The patient made a good recovery, but is now suffering from slight menopausal symptoms.
>   [But] when I came to examine the sections myself, I came to the

conclusion that the left genital gland was an ovo-testis, and not the seat of a malignant growth as reported [by the pathological laboratory].[67]

Blair Bell had found an ovotestis in a living patient (see Figure 22). But Blair Bell did not believe that S.B.'s case should be considered one of "true hermaphroditism." He noted (like Pozzi) that in cases like S.B.'s, where there was testicular and ovarian tissue in the form of an ovotestis, the subjects presented "predominating feminine characteristics [which] have [invariably] decided the sex adopted."[68] (A curious claim in any case, given the bodily features of S.B. as shown in Figure 21 and as described in Blair Bell's text.) It therefore seemed to Blair Bell that the presence of ovotestes would better be designated—following the recommendation of Tuffier and Lapointe, two French investigators—"partial glandular hermaphroditism," partial because it did not result in the "possession of *all* the genital organs of both sexes, with the possibility of fertilising themselves, or at least with the power of fertilising others and of being fertilised."[69]

The issue was not just one of gland functionality, though. With biopsies now becoming, in the 1910s, standard practice in cases of doubtful sex, the presence of ovotestes could be—and increasingly would be—confirmed in living patients. What was one supposed to do with regard to sex assignment when faced with a living person with an ovotestis? Now it seemed to make more sense practically as well as theoretically to consider a case such as S.B.'s one of "partial glandular hermaphroditism in a *woman*" rather than a one of "true hermaphroditism." This way there would still be only two sexes on the streets, as well as in the books. Tuffier and Lapointe made the point explicitly, and Blair Bell implicitly agreed: "The possession of a [single] sex is a necessity of our social order, for hermaphrodites as well as for normal subjects."[70] S.B. and other true hermaphrodites were now to be women suffering from "partial glandular hermaphroditism."

The second case that strongly influenced Blair Bell to consider revision of the anatomical-gonadal definition of sex had been uncovered by Dr. Russell Andrews of London. Andrews's subject was twenty-three years old:

'She' was very good-looking, and had the voice and all the secondary characteristics of a woman, for which 'she' had always passed. 'She' sought advice for primary amenorrhoea and abdominal pain. On ex-

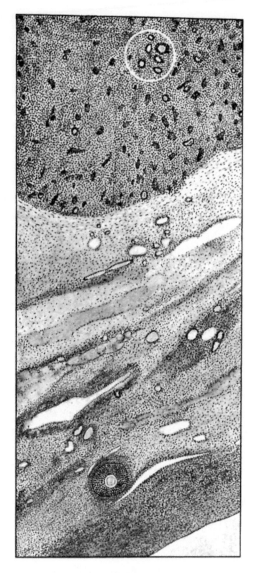

Figure 22 Photograph of a cross-section of the left genital gland of S.B., said by Blair Bell to be an ovotestis—the top section testicular, the bottom ovarian. The white circle at top highlights alleged seminiferous tubules (testicular features).

amination, the external genitalia were found to be of the normal feminine type; there was a short vagina (uro-genital sinus) about an inch and a half in length and of normal diameter, but no uterus could be felt.

In these circumstances, but not doubting 'her' sex, Dr. Andrews thought it advisable to open the abdomen; he then found there was no uterus and that there were two non-functional testes with vasa deferentia. This woman was a man, or rather an external partial hermaphrodite.[71]

According to the now long-standing tradition of sex diagnosis that took the anatomy of the gonad as the marker of essential sex, Andrews's patient would indeed be a man. But this seemed to Blair Bell absurd, theoretically and practically. The testes were "non-functional" (Blair Bell simply presumed this because she looked so womanly); why should they mark true sex? And, a more pressing question still, what purpose would it serve—or what havoc would it wreak—to declare a person like the one described above and shown in Figure 23 a man?

In Blair Bell's opinion, to assign sex according to the gonadal anatomy alone was to risk not only the happiness of the patient, but social order as well:

> I want . . . to raise the question as to whether we are justified . . . in branding [patients] with a sex which is often foreign not only to their appearance but also to their instincts and social happiness. In the case of the male glandular partial hermaphrodite recorded by Tuffier and Lapointe [as in the case of S.B.] the appearance and instincts were feminine, even if some of the external and internal genital characteristics were masculine. In the case of Russell Andrews there was absolutely nothing masculine about the patient except non-functional testes. Both these patients were good-looking and might have been married, just as many other male tubular partial hermaphrodites have been, believing themselves to be women.[72]

Blair Bell continued:

> Since it is now possible to demonstrate the fact that the psychical and physical attributes of sex are not necessarily dependent on the gonads, I think that each case should be considered as a whole; that is

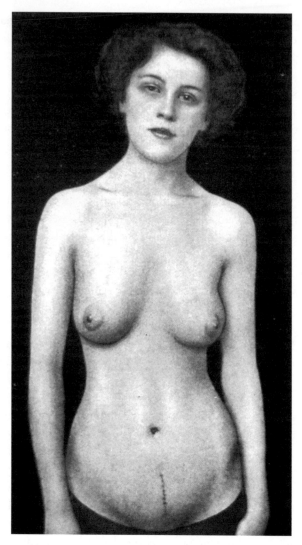

Figure 23 Russell Andrew's patient, who had two testicles and no ovaries. Note her laparotomy (exploratory surgery) scar.

to say, *the sex should be determined by the obvious predominance of characteristics, especially the secondary,* and not by the non-functional sex-glands alone, for this is neither scientific nor just.[73]

Blair Bell was not, of course, suggesting that—just because the gonad was not "the only ruling factor" in "the development and maintenance" of sex—that sexual characteristics and sex differentiation originated from some immaterial basis.[74] It might be true that "most men, perhaps, are content to believe that women are the gifts of the gods while women probably look upon men as inventions of the devil."[75] Blair Bell confessed with some frustration that "judging by some of the remarks I have heard at meetings—supposed to be scientific—when this subject has been under discussion, I fear the average mind looks upon all primary causes as beyond scientific investigation."[76]

But Blair Bell thought the matter of sex differentiation quite open to study, and indeed he believed much of the problem was already sketched out thanks to investigations into abnormal sexual differentiation:

> We know from the researches of Bulloch and Sequeira, which have been confirmed and extended by Glynn and others, that hyperplasia and certain neoplasms of the suprarenal cortex may produce definite masculinity in women previously feminine. So, too, I have shown that acromegaly, due to hyperplasia of the anterior lobe of the pituitary, produces many of the symptoms of masculinity, including amenorrhoea, enlargement of the clitoris, deep voice, and a thick, coarse skin.
>
> It appears probable, then, that all the organs of internal secretion may have a certain definite influence—stimulating or restraining—on the development of special sex-characteristics, both primary and secondary.[77]

So, Blair Bell concluded, when so many glands could contribute to the "development of special sex-characteristics," what sense did it make to privilege some nonfunctional testes or ovaries as markers of true sex? And what sense did it make to say that someone was a true hermaphrodite when we had no way to deal with that socially or legally?

It is important to note that, in spite of his revolutionary prescription for sex diagnosis, one which moved overtly and consciously away from the gonadal definition, Blair Bell's position was conservative in two fundamental ways. First, Blair Bell was, like his predecessors and successors, motivated in theory and practice by an interest in maintaining clear, medically sanctioned divisions between the two sexes in each individual case and in society as a whole. Indeed, this was largely the reason Blair Bell suggested the abandonment of the gonad-as-exclusive-marker rule. That rule, now obsolete because of new diagnostic technologies, would have left S.B. a living true hermaphrodite, and Andrews's patient an unbelievably feminine man. S.B. and Andrews's patient looked like, acted like, thought themselves to be, and were thought to be, women. What sort of ramifications might result from announcing that they were men or truly hermaphroditic? If men and women were to be kept distinct, Blair Bell realized, hermaphrodite-sorting would have to be accomplished in such a way as to quiet sex anomalies, not accentuate them.

Second, Blair Bell still maintained the idea that every body did indeed have a single true sex, even if "neither the sex-gland nor the genital ducts necessarily influence or give any indication of the true sex of the individual, as shown by the secondary characteristics."[78] Blair Bell recommended that medical doctors not only diagnose a single sex for anomalous bodies, but that they then help it along, by eliminating any sexually "anomalous" characteristics and accenting those that matched the so-diagnosed sex. In Russell Andrews's case, for instance, "the patient was undoubtedly an attractive woman, unfortunately with testes. [But] these might have been entirely removed. Such a procedure would have left 'her' a woman, for all intents and purposes except conception."[79] In a paragraph containing the first use I find in hermaphrodite literature of the word "gender," Blair Bell concluded:

> [O]ur opinion of the gender [of a given patient] should be adapted to the peculiar circumstances and to our modern knowledge of the complexity of sex, and . . . surgical procedures should in these special cases be carried out to establish more completely the obvious sex of the individual.[80]

So true sex would, perhaps, no longer be dictated exclusively by the anatomical nature of the gonads. But only two true sexes would still exist, with a limit of one to each body, and the medical man would still be the interpreter—and now, when necessary and possible, the amplifier—of true sex.

This—the assignment to and the surgical construction of a single, believable sex for each ambiguous body—was the way of the future. Blair Bell's work sounded the end of the Age of Gonads, but it harkened in a successor era in which, as before, each body would be allowed only a single true sex, and the medical doctor would be the determiner or even the creator of it. Sex would now be consciously and literally constructed as certain whenever it appeared dangerously doubtful. Medicine had won out over the threat of the hermaphrodite.

EPILOGUE

Categorical Imperatives

One of our most difficult duties as human beings is to listen to the voices of those who suffer . . . These voices bespeak conditions of embodiment that most of us would rather forget our vulnerability to. Listening is hard, but it is also a fundamental moral act; to realize the best potential in postmodern times requires an ethics of listening.

—Arthur Frank, *The Wounded Storyteller* (1995)

[G]enital ambiguity is "corrected" not because it is threatening to the infant's life but because it is threatening to the infant's culture.

—Suzanne J. Kessler, "The Medical Construction of Gender"

THIS BOOK WOULD LOOK VERY DIFFERENT if it could have included first-hand accounts telling us how people labeled "hermaphrodites" in the nineteenth century saw and represented their own bodies and lives. Unfortunately, as the septuagenarian British hemophiliac Donald Bateman noted in a 1994 autobiographical article, "the social history of medicine is usually recorded by its practitioners, by social workers, or researchers. Not much of it is chronicled by its victims or the recipients of treatment. The sick, like the poor, leave very few archives behind them."[1] This has been equally the case for those not necessarily "sick" but nonetheless labeled anatomically abnormal and therefore subject to medical attention: hermaphrodites, like the poor and sick, have left few personal archives. In earlier chapters I documented the rise of medical authority and, by consequence, the general usurpation or miscarriage of authoritative voice that subjects of medicine may have otherwise had.

The late twentieth century, however, has seen the emergence of the voices and claims—to autonomy, to authority—of medicine's subjects. Intersexuals, like hemophiliacs and other medical patients, have begun to record and make known their stories in ever greater numbers. There is significant value in listening to intersexuals' autobiographies. As in the personal histories of "interracial" people, in intersexuals' stories we can hear first-hand what it is like to live on one of the great cultural divides. These individuals elucidate not only the life of the ambiguous and the ambiguities of life but also the many ways anatomical realities and identities are created, divided, and restricted in culture. More important still, by paying attention to the autobiographies and thoughts of intersexuals, we see the problems inherent in the way U.S. medicine and culture deal with intersexuality. Recent hermaphrodite autobiographies almost unanimously challenge medical doctors today to give up the typical treatment of them—a treatment rooted in a medicine and a culture of old—and challenge the rest of us to recognize their existences and problems.

This epilogue briefly discusses three issues: (1) the present-day medical treatment protocols for intersexuality, which call for the creation, as soon after birth as possible, of a "believable" masculine or feminine anatomy via plastic surgery and hormonal therapy, and a silencing of any doubt parents or others might have; (2) how those protocols, however well intentioned, maintain many vestiges of nineteenth-century medical theory and practice; and (3) the present moment in which intersexuals are finally themselves challenging medical treatment protocols and the rigid cultural categories that impel those protocols.

Of late, medical institutions have been subject to many challenges from inside and outside their ranks. In order to understand today's medicine and its problems and challenges, we must trace its various histories. Only with a close study of the history of medical approaches to hermaphroditism can we understand why doctors and intersexuals do what they do today and what options there might be.

In taking very seriously the autobiographies of intersexuals today, this epilogue also places this book in the realm of what has come to be known as "narrative ethics," that is, medical ethics philosophies that take into account the substance and nature of patients' stories. Indeed, one of this chapter's epigraphs is from a text on narrative ethics. In *The Wounded Storyteller: Body, Illness, and Ethics,* Arthur Frank explores the narratives created in and around suffering, especially autobiographical narratives of presently ill and formerly ill people and their caregivers. Frank and I both seek to draw historical and sociological lessons from these accounts and, moreover, to understand the way in which narratives of suffering carry with them ethical obligations—obligations on the part of sufferers to tell their stories, and obligations on the part of others to listen with the storyteller and to recognize the community (and the communal obligations) of the culture in which these narratives are created.[2]

The phrase "listening *with*" (rather than *to*) is unfamiliar, because we are accustomed to the idea that we listen passively to a story and then draw what conclusions we may for ourselves. But Frank insists that we in fact do more through narratives of suffering. He demonstrates, for instance, that, as a storyteller tells and repeats a story, she or he also embodies and amplifies larger narratives (mythologies) and narrative structures (tropes) of the larger culture. Some wounded storytellers, for example, tell "quest" narratives that follow closely in form and purpose the ur-quest mythology that Joseph Campbell presented in his well-known book *The Hero with a Thousand Faces*.[3] A wounded storyteller—or a doctor recounting an intersexual's "case history" to colleagues, or a historian summarizing intersexual narratives—in a sense is often "listening with" and weaving in the other, larger cultural stories as she tells her individual story. For example, as we have seen, medical case reports of the nineteenth century often contained culturally dominant stories about the proper roles for the sexes and for the patient and doctor. (We will see below that this is still the trend in medical discourse on intersexuality.) Similarly, the stories of hermaphrodites' mothers often were built around the trope of maternal guilt for an "imperfect" or "marked" child. And I have aligned my historical narrative with larger stories about medical and bodily authority. To understand the stories we all tell, and to understand the actions taken as a result of those stories, we need to recognize that we "listen with" surrounding cultural narratives. A story is never an island unto itself; if it were, it would be almost incomprehensible and therefore not really recognizable as a cohesive story.

Frank also suggests another norm for "listening with" a wounded storyteller: listening with and "[t]hinking with stories means joining with them; allowing one's own thoughts to adopt the story's immanent logic of causality, its temporality, and its narrative tensions . . . The goal is empathy, not as internalizing the feelings of the other, but as what Halpern calls 'resonance' with the other."[4] In this act of listening, the storyteller and her fellow listeners become valuable witnesses to the world and nature of suffering, disease, medicine, disability, and so on. Ethical behavior means recognizing and respecting the imperatives embedded in stories of suffering. Recognition of these obligations and attention to them, Frank argues, form the basis for a narrative ethics of medicine, an ethics that "[u]ltimately . . . [recognizes] how much we as fellow-humans have to do with each other."[5]

Postmodernist Intersexuality

Why are intersexual autobiographies suddenly appearing in relatively large numbers? We could chalk up the emergence of a discourse by and among intersexuals to the electronic-media technological revolution. The Internet, the World Wide Web, and the pervasiveness of other media (including television) have made intersexuals known and known to one another, and of late, via these media, intersexuals have often inspired one another to recognize their common problems and tell their individual stories. A simple technological determinism, however, does not suffice to explain the phenomenon. The bloom of intersexual autobiographies is also traceable, I think, to the impetuses of postmodernism.

Postmodern times have enabled the emergence of intersexual autobiographies in five ways. First, postmodernism has seen the valuing of voices previously considered nonauthoritative.[6] The hemophiliac Bateman's autobiographical article was, for instance, published in an academic journal that in the past would have been extremely unlikely to publish such a text by a technical "nonexpert" in medicine or medical history. In these postmodern times, however, Bateman is recognized as something of an expert. The journal's editor chose to label Bateman's piece a "subjective chronicle," as if to suggest that a purely objective chronicle of hemophilia is possible, yet scholars have acquired enough doubt about beliefs in simple objectivity to decide that Bateman's chronicle—however "subjective"—is worth recording and disseminating. Previously, such autobiographies would only have been considered "primary sources" (that is, sources in need of selective analysis) for medical theorists and medical historians. Today, like people living with AIDS, hemophilia, schizophrenia, or cancer, people born intersexual consider themselves important enough to speak and to be heard, and they are slowly becoming authorities of sorts.

Second, postmodernism has brought with it the recognition that there can never be a single, self-evident, "true" story to be told about a life, disease, or condition. In the past, if a relatively disempowered person's story conflicted with the dominant story, the socially weaker individual's tale was likely to go unheard or discounted. Postmodernists like Frank recognize, however, that the decision to call one story "true" or primary is a complicated one involving many value choices. With this recognition, intersexuals and others are now able to tell and "hear" stories about intersexual life that challenge or conflict with the classic modernist medical story about relatively simple, containable, attacking invader "diseases" and heroic, purgative "cures."[7] Postmodernist intersexuals refuse to take their doctors' stories about them, their "problems," and their treatments as primary; instead they reject, change, or incorporate medical narratives into their own narratives and make their own stories primary—at least in their own lives.

Third, postmodernist sufferers often share a sense that their bodies have been "colonized" by medicine in ways that impel them to resist and object. Frank writes that "becoming a victim of medicine is a recurring theme in [postmodern] illness stories" like Bateman's. Frank notes that, today, "often physicians are understood as fronting a bureaucratic administrative system that colonizes the body by making it into its 'case.' People feel victimized when decisions about them are made by strangers. The sick role is no longer understood as a release from normal obligations; instead it becomes a vulnerability to extended institutional colonialization."[8] Feminists (like the authors of *Our Bodies, Ourselves* and *The New Our Bodies, Ourselves*), sufferers of depression, and many others have in the last few decades objected to the seizing and dividing up of their bodies by doctors and scientists in ways that, rather than restoring strength to them, seem to sap them of power.

Like many recipients of intensive medicine, then, intersexuals express mixed or ambivalent feelings toward medicine. Frank confesses:

Some of my deepest, even haunted, discussions with other members of the remission society [people who have "come through" an illness, at least temporarily] have been attempts to sort out whether chemotherapy is a form of torture. We know that in most "objective" respects the two situations differ, and we seek only to make sense of our own memories and fears, not to appropriate the far greater suffering of torture victims. But chemotherapy fits with disturbing ease into Elaine Scarry's definition of torture as "unmaking the world." . . . [In chemotherapy] I was unmade as my mind sought to hold onto the promise that this treatment was curing me, while my body deteriorated: my intactness, my integrity as a body-self, disintegrated . . . Chemotherapy is hardly the only occasion for comparing medical treatment to torture.[9]

In the treatment of intersexuality as in the use of chemotherapy, the physician sees the treatment as necessary, but, even when the patient agrees with the necessity of the treatment (which not all intersexuals do), the recipient of treatment paradoxically feels destroyed, divided, and "unmade" by the very process aimed at healing.

Thus, fourth, the modernist conception of the active physician-hero—a strictly rationalistic, brave, selfless savior who treats a silent, passive, unambiguously grateful patient—has given way to postmodernist challenges of the doctor-patient balance of power and to challenges to the "doctor as savior" motif. Webster's *Ninth New Collegiate Dictionary* defines "patient" as "an individual awaiting or under medical care and treatment; one that is acted upon." In other words, the patient has traditionally been not an actor but rather a subject who is acted upon. By contrast, patients are now often termed "health-care consumers," a designation that signals their new role as actors (which is not to say the newer term is not problematic in its own right). The shift toward more active "patients" has occurred in part because recipients of medicine have sought more autonomy and authority, as have medical ethicists on their behalf; intersexuals are not alone in their growing confidence to ask difficult questions and demand the right to make key choices in their treatment. But the shift in roles has also occurred because many physicians have asked patients to take a more active role in their health care and to understand that, first, doctors are human and that, second, medicine cannot cure everything and does not work identically in every case.[10] The image of the doctor as superhuman, almost divine (in intent and/or talent) is under scrutiny from all sides.

Finally, postmodernism, in its appreciation of the social construction of concepts like sexual identity and normality, has given intersexuals the opportunity to see their plight as contingent to social times and places—to see their experiences as culturally, historically specific and therefore not inherent in or necessary to their bodies. Cultural histories such as the one presented in this book have demonstrated the cultural dependency of the categorization and treatments of males, females, and hermaphrodites. Awareness of the "social construction" of these categories has enabled intersexuals to object to their treatment as "freaks" or "problems" to be corrected and disappeared.[11] Intersexuals have also looked beyond the history of hermaphrodites to the history of the medical treatment of homosexuality and have noted that, until very recently, gay and lesbian people were also widely treated as troubling and troubled "freaks" to be "fixed."[12] The history of the biomedical construction of womanhood reads similarly.[13] Intersexuals have realized that, like straight women and gay people, they need not be treated as fundamentally unacceptable or flawed—that it is not their bodies that make their lives difficult, but the cultural demands forced upon their bodies. Many of them hope and believe it might be possible to construct a world that would not reject them or insist that they be uniformly and exclusively treated by medicine.

So, for the first time, numbers of intersexual autobiographies are emerging to challenge the physical, theoretic, and linguistic treatment of their bodies by others. Some of these autobiographies are extremely critical of the medical treatment of intersexuals; others are less critical; some are largely narrative; others are deeply theoretical; but all tell a story more complex than can be told by any medical, historical, or theoretical text.

DIANE MARIE ANGER has been diagnosed as a male pseudohermaphrodite, although it took her a long time to find this out. She was born in the late 1960s with testes and with an XY chromosomal basis, but apparently at birth her external genitalia were such that the doctors attending her case decided she would most successfully be made a girl. (The logic for these decisions is discussed below.) At three years of age, Anger's testes were removed, and when she was eighteen, she underwent surgery for vaginoplasty (construction of a "vagina"), urethral reconstruction (to move her urethral opening to a more "typical" place), and breast implantation. Anger was relieved to have the surgery,

> because all of my life the ambiguous genitalia . . . [were] a constant reminder of my freakish nature and a real fort [*sic*] of fear and anxiety and social isolation from the rest of the world. I thought I would wake up with a beard and a deep voice and that I would grow a full fledged penis! No one had ever bothered to tell me that [that]

couldn't possibly happen because my "gonad[s]" [had been] taken out. I only discovered that when I had vaginoplasty done at 18. And then my doctors still weren't willing to divulge much info[rmation], nor were my parents (my dad was rather embarrassed about the whole process).

Anger has suffered from depression since early childhood (she has attempted suicide three times), and she had thought that the series of "reconstructive" operations would finally make her look and feel the way she thought other women looked and felt. The attending physicians assured her parents that all would be well—that is, that she would have a "normal," "feminine" (meaning heterosexual) life: Anger recalls her parents' being told, "Don't worry, another patient with this [condition] is getting married now."

Unfortunately, the operations did not go as well as the doctors had intended, and three years later, Anger found her persistent fears realized when the physician who had performed the vaginoplasty confessed to her, "I'm disappointed it didn't turn out as I had hoped." The "vagina" was placed "too far down" with the result that, not only does it not look convincing, but whenever Anger has bouts of diarrhea, the diarrhea "splashes up" into the "vagina." Anger's ability to enjoy orgasms has been greatly reduced by the surgeries and their so-called side effects, and she cannot feel anything in the "vagina" that was built for her.

Anger's urethral reconstruction has also been a source of serious problems. The body's self-preserving reaction to wounds is to heal, and part of that healing involves growing repair tissue. Unfortunately, this often means the build-up of significant scar tissue after surgery. Scar tissue in the urethra often leads to infection, and about three years after her surgeries, Anger began to suffer from a series of urinary tract infections that would last over seven years. To date she has had approximately seven surgeries to reopen her urethra and to repair a bladder neck contracture (which caused discomfort before and after the surgery to fix it). Having been through so many surgeries and the associated physical and emotional trauma, Anger finally decided it would be better to dilate her urethra herself once a month, or more often if necessary. She continues to do self-dilations out of fear of additional surgery.

Like the physician who performed the vaginoplasty, Anger finds the reconstruction unconvincing, and she worries that lovers may not want to be with her. From childhood, the vague hints and innuendoes Anger had picked up about her anatomy had left her confused about who she was and who she should be. "I grew up wondering if I should like boys or girls because if I like[d] boys then I thought I was a homosexual because I was genetically XY and had male reproductive organs removed, [but] if I like[d] girls I thought I was still a homosexual because I was raised a girl." Anger "wishes that counseling had been offered" to her from early childhood, because such assistance might have forestalled her depression and the "emotional turmoil" she has experienced "over dating, sex, marriage, and sexual identity."

After two years of therapy and medication for her depression, Anger feels that she has come to accept her condition and accept herself "as a valued, contributing person." She also has benefited from finding a urologist who takes her concerns very seriously. Anger notes that she holds a degree in biology and has been a medical writer and researcher in the field of orthopedic surgery for the last five years. She says that she has "learned to be independent and ambitious because I realize the chance of finding a life-long mate is low," and in any case, she "isn't so keen on sharing my life at this point with anyone, except my furry friends." Anger now lives in Florida with her two pet dogs and six pet guinea-pigs. She also acts as a foster mother for animals; when I heard from her, she was raising, in addition to her own eight pets, another "three puppies and their mother, all of whom have severe mange, for the Florida Humane Society."[14] The photograph of herself she offered for this book provides a notable contrast to medical representations of people born intersexual (see Figure 24).

"One of the things about being born with genitals that challenge what is considered normal," writes Martha Coventry, "is that no one ever tells you that there is anyone like you. . . . You are purposefully isolated, your difference covered up, and it is horrible." Coventry was born in the early 1950s a healthy little girl. Her clitoris happened to be a bit bigger than most—it peeked out slightly beyond her labial folds. Coventry does not remember as a child thinking anything particular about her clitoris, until suddenly, when she was six years old, she was brought to the hospital for an operation, one which was meant to end all questions about her genitalia. The surgery may have ultimately forestalled questions for others who would know her. But for Coventry, it began decades of confusion, denial, and doubt.

Coventry's clitoris was "resected," in the medical language of surgery—cut out, shortened, made almost invisible. When she was eleven or twelve, roughly six years after the operation, Coventry came upon her

Figure 24 Diane Marie Anger (1997), as she likes to be seen.

parents looking at baby pictures. "My mother picked mine up and I heard the word 'boy' come out of her mouth. Fear heaved in me. I was a boy. I was supposed to be a boy." Unable to stop herself, Coventry blurted out, "What was that operation I had?" The only answer came from her father, a man whom she loved deeply and a man who was also an established surgeon: "Don't be so self-examining." And then, for Coventry, "the moment of silence that followed that brusque dismissal lasted for almost twenty-five years." Only in the last few years has she undone the silence. A new gynecologist, "a wise, irreverent man," has worked with her to try to understand what her genitalia probably looked like before the surgery. Coventry now writes as a form of therapy, and her prose bursts with an intense celebration of her sexuality and her being.[15]

Most of the intersexuals whose stories I have located are members of the Intersex Society of North America, a group founded by Cheryl Chase.

I will always remember the evening we met because she had, just the day before, received from her aunt copies of her baby pictures dating from when she had been a boy in her family. Consequently I was looking at these photographs only hours after the first time she herself had seen them. The doctors who advised Chase's parents about her sex reassignment told them to get rid of all of the testaments to her boyhood—the name "Charlie," the baby blue clothing, the birthday cards, and the pictures—and to make no mention of what had been. For whatever reason, Chase's aunt kept her copies of the pre-reassignment baby pictures, and she gave them to Chase when Chase asked for them in 1996.

Cheryl Chase was born Charlie in the late 1950s, and it was not until she reached her early twenties that she began to reclaim her history. Frank notes that "in postmodern times 'reclaiming' has been used to the point of cliché, but like most clichés it carries a significant kernel of truth about the times in which it is so often repeated. Reclaiming suggests that illness stories are doing more than speaking through interruptions; the ill [or suffering] person's voice has been taken away."[16] Chase is still reclaiming her history, collecting those bits of stories, voices, and pictures that escaped the purge. She writes of her struggle to find out what happened:

> It took months for me to obtain . . . [all] of my medical records. I learned that I had been born, not with a penis, but with intersexual organs: a typical vagina and outer labia, female urethra, and a very large clitoris. Mind you, "large" and "small," as applied to intersexual genitals, are judgments which exist only in the mind of the beholder. From my birth until the surgery, while I was [considered a boy], my parents and doctors considered my penis to be monstrously small, as well as lacking a urethra . . . Then, in the moment that intersex specialist physicians pronounced that my "true sex" was female, my clitoris was suddenly monstrously large.[17]

And so it was cut off. The medical record stated succinctly: "Diagnosis: true hermaphrodite. Operation: clitorectomy." Charlie's name was changed to Cheryl, and her past was mostly discarded along with her clitoris. Chase remembers her parents entrusting her with the vague details of the operation when she was ten: "When you were a baby," they explained, "you were sick . . . Your clitoris was too big, it was *enlarged* . . . You had to go into the hospital, and it was removed . . . Now everything is fine. But don't ever tell this to anyone else."

As an adult, Chase has opted to do quite the opposite—to tell many people, including many strangers, about her experience, because she believes that what happened to her should not happen to anyone else. "The time has come," she writes in her article "Affronting Reason," "for intersexuals to denounce our treatment as abuse, to embrace and openly assert our identities as intersexuals, to intentionally affront that sort of reason which requires that we be mutilated and silenced." Like many self-acknowledged intersexuals, rather than denying the paradoxes of being an intersexual, Chase defiantly embraces them: "I claim lesbian identity because it is lesbian women who feel desire for me, because I feel most female when being sexual, and because I feel desire for women as I do not for men. Most intersexuals share my sense of queer identity, even those who do not share this homosexual identity." "Queer" is how many come to feel, given the frenzied silences directed at their genitals.

Sven Nicholson, now in his mid-forties, was born with a hypospadic penis. Recall from Chapter 1 that hypospadias is the name for the condition in which the urethral opening is located somewhere other than the very tip of the penis. Many doctors, particularly urologists, consider hypospadias a problem because a hypospadic penis can look unusual or require a boy or man to urinate sitting down. Therefore hypospadic penises are often surgically reconstructed in an attempt to move the urethral opening to the "right" place. "When I was eleven," Nicholson recalls,

> I had three operations to repair [the] hypospadias. These operations were performed by a competent physician who considered my family a charity case and never sent us a bill. The artistry of his work has been commented on by most urologists who have subsequently examined me. He sincerely believed this was the best treatment for me, and did the best job he could. However, a stricture developed within two months of the final operation, and ever since my life has been drastically altered.[18]

"Stricture" is the name for a closing-up; Nicholson's stricture resulted from scarring in the urethra caused by the "corrective" surgery, and it left him facing dangerous and debilitating urinary tract infections. For the remainder of his life, frequent catheterization became necessary. In this procedure a tube is inserted up the urethra toward the bladder. The goal is to re-open the closing urethra and to drain the bladder of urine before infections set in.

Because of the problems caused by the surgery, Nicholson wound up traveling weekly to a series of physicians who would use a variety of techniques to catheterize him. Some of the techniques were less painful than others, but all were extremely unpleasant at best. "Throughout my teenage years, physicians seemed to take the attitude that my condition could be somehow healed through a catheterization regime. In my early twenties, the physicians dropped this pretense" and essentially admitted to Nicholson that he would have to undergo catheterization every few days for the rest of his life. He might have continued going to physicians for this procedure, had it not been for a vocation that took him away from ready access to doctors:

When I finished college, I became a lay missionary for my church. The mission board physician asked my urologist for a letter stating that my condition would not cause problems overseas. The physician gave me a set of silicon-coated catheters and instructed me in their use. His casual attitude reassured me, "You can live anywhere in the world, as long as there's soap and warm water." However, he was extremely reluctant to commit this to writing; in retrospect, I think that he did not want to create any written statement that a medical condition such as mine can be casually and easily treated by the patient himself. In any case, he wrote the letter at the last minute (which I never saw) and gave me a generous supply of antibacterial sulfa drugs and anesthetic lubricant, and I was on my way.

For several years, Nicholson continued to rely on physicians for prescription antibacterial drugs, but when, in his mid-thirties, he "visited a urologist who refused to prescribe these drugs unless he first performed another surgery on me, to the tune of several thousand dollars[,] I became determined to learn how to open my urethra without any prescription drugs." He did so, and Nicholson now considers "self-catheterization a vast improvement over the visits to the urologist's office." Indeed, he has written a short autobiography which includes a step-by-step "guide to home auto-catheterization." Catheterization, he notes, "is a type of personal grooming. I expect as much pain and pleasure from it as I do from shaving or brushing my teeth."

Nicholson has many questions about the logic and behavior of the various physicians he has encountered: "In retrospect, I wish that the operations had never happened, that I had simply been allowed to live out

my life with the plumbing system originally given to me by my Creator."
Although the original series of operations, designed to make his penis
look more "normal," was explained briefly to Nicholson before it began,
"alternatives were never discussed, no scenario other than the desired
outcome was ever presented. I had never heard of 'informed consent.'"
Even though he recalls the world of his youth being "full of references to
Freud, to counselors and therapists and psychoanalysts and pastors who
had a counseling ministry, etc. etc. etc.," psychological counseling was
never offered him or his parents. "And probably my doctors never looked
at the problem in psychological terms . . . I was encouraged to deny
reality, to think that the cure lay in just a few more visits to the doctor."

Nicholson now trusts himself as an authority on the matter of catheterization, and he finds himself, as something of an expert, wondering why it
is that several of the urologists he encountered insisted that he must dilate
his urethra to an extreme width, something he finds unnecessarily painful.
"When I was visiting physicians during my teenage years, a major emphasis was placed on dilation up to [a size] 24 Foley [catheter]. I'm not sure
that my masculinity depends on the internal diameter of my urethra; in
terms of plumbing, anything above 14 is fine."

Modernist Medical Protocols

The life histories recounted by Diane Marie Anger, Martha Coventry,
Cheryl Chase, Sven Nicholson, and other intersexuals tend to be postmodernistic: they describe colonized bodies and lives; they reject many of the
claims laid upon them by others and see medicine as one deeply troubling
aspect of their troubled lives; and they seek to be heard as authoritative
despite their historically unauthoritative position. They also, consciously
and unconsciously, refuse to constrict their stories to simple tropes or to
relinquish themselves to simple categorization. For example, the story of
Diane Marie Anger contains a mix of appreciation and loathing for surgery.
Martha Coventry enjoys the possibilities of an "intersexual" identity even
though the removal of her "oversized" clitoris—her only questioned
part—was also supposed to have removed any possibility of that identity.[19]
Cheryl Chase refuses to have her identity labels chosen for her, or even to
have them completely stabilized. And Sven Nicholson challenges the way
doctors have attempted to take control of and construct his manhood and
his identity as a "patient in need of medical attention."

By contrast, present-day medical discourse and the medical-technological approaches to intersexuality are extremely modernistic. Typically, after the hospital birth of an intersexed baby, teams of specialists (geneticists, endocrinologists, pediatric urologists, and so on) are immediately assembled. These doctors decide to which sex/gender a given child will be assigned, and the power of modern medicine is then brought to bear to create and maintain that sex in as believable a form as possible. Three kinds of activity characterize this sex creation: surgery on the genitals, and sometimes later also on other "anomalous" parts like enlarged breasts in an assigned-male or unenlarged breasts in an assigned-female; hormone monitoring and treatments to find a combination that will help and not contradict the decided sex (and that will avoid metabolic dangers); and, with the presumption that the first two activities are required for the third to follow, the subsequent psychosocial rearing of the child according to the norms of the chosen sex.[20]

Intersexuality experts worry that any "confusion about the sexual identity" of the child on the part of relatives will "transmit to the child" and result in enormous psychological problems for the child, including potential "dysphoric" states in adolescence and adulthood. Among the "dysphoric" states doctors have traditionally worried about is homosexuality, understood by intersex doctors as when a person born intersexed has sexual relations with people of the same assigned gender. Dysphoria is defined as the "state of feeling unwell or unhappy"; homosexuality in this schema is therefore labeled an unsuccessful and unfortunate way of being.[21] In an effort to forestall or end any confusion about the child's sexual identity, doctors try to see to it that an intersexual's sex/gender identity is permanently decided by specialist doctors within forty-eight hours of birth. With the same goals in mind, intersex doctors insist that parents of intersexed newborns be told that their ambiguous child *does* have a male or female sex, but that the sex of their child has not yet "finished" developing, and that the doctors will quickly figure out the "correct" sex and then help surgically "finish" the sexual development.[22]

Unlike doctors of the late nineteenth century, those of the late twentieth century do not search deep into an intersexual's body in the hopes of finding a material marker of an ontologically "true" sex. Instead, doctors today see their approach as pragmatic and primarily attentive to a psychosocial gender-identity theory rather than a biomedical-materialist philosophy of sex identity. That psychosocial gender-identity theory, established by John Money in the 1950s, holds that all children must have their gender identity fixed very early in life; that from very early in life children's anatomy must match the "standard" anatomy for their gender; and that boys primarily require "adequate" penises with no vagina, while girls primarily require a vagina with no easily noticeable phallus.[23] This psychosocial theory of intersexuality presumes that these rules must be followed if intersexual children are to achieve successful adjustment appropriate to their assigned gender—that is, if they are to act like girls, boys, men, and women are "supposed" to (including heterosexually and not homosexually). The theory then also by implication presumes that there are definite acceptable and nonacceptable roles for boys, girls, men, and women, and that this approach will achieve successful psychosocial adjustment, at least far more often than any other approach.

Because of widespread acceptance of this theory, the practical rules now adopted by the majority of intersex specialists are these: genetic males (children shown to have Y chromosomes) must have acceptable penises if they are to be assigned the male gender. If their penises are determined to be "inadequate" for successful adjustment as males, they are assigned the female gender and reconstructed to look female. In cases of intersexed children assigned the female sex/gender, surgeons refashion phalluses to look like clitorises (or at least to be invisible when the individual is standing), build vulvas and vaginas if necessary, and remove any testes. This is done even if it means risking a child's only real chance at becoming a biological parent, because intersex doctors consider "adequate" penises far more important for boys than potential fertility.[24] These children often also receive "feminizing" hormonal supplements and later breast implants to accentuate "feminine" development. This was the case, for instance, for Diane Marie Anger, who was diagnosed a male pseudo-hermaphrodite. Genetic (XY) males with "adequate" phallus size may have their penises reconstructed to look normal; this, for example, is the case with hypospadic penises and was the case for Sven Nicholson.

Meanwhile, genetic females (that is, babies lacking a Y chromosome) born with ambiguous genitalia are declared girls—no matter how masculine their genitalia look. This is done chiefly in the interest of preserving these children's feminine reproductive capabilities and in bringing their anatomical appearance and physiological capabilities into line with that reproductive role. Consequently, these children are reconstructed to look female using the same general techniques as those used on genetic-XY children assigned a female role. Surgeons reduce "enlarged clitorises" (like Martha Coventry's) so that they will not look or act "masculine."

Vaginas are built or lengthened if necessary, in order to make them big enough to accept average-sized penises. Joined labia are separated, and various other surgical and hormonal treatments are directed at producing a believable and, it is hoped, fertile girl. Clitorises—meaning phalluses in children assigned female identities—are typically considered too big if they exceed one centimeter (0.39 inches) in length.[25] Intersex-specialist pediatric surgeons consider "enlarged" clitorises to be "cosmetically offensive"[26] in girls and therefore they subject them to surgical reduction meant to leave the organs looking more "feminine" and "delicate."[27] Penises—meaning phalluses in children assigned male identities—are typically considered too small if the stretched length is less than 2.5 centimeters (about an inch). Consequently XY children born at term "with a stretched penile length less than 2.5 [centimeters] are usually given a female sex assignment."[28] (Against the 2.5-centimeter-grain, at least one doctor has set the acceptable lower limit at 1.5 centimeters; such a decision makes clear the socially constructed nature of these measurements.)[29]

Roughly the same protocols are applied to cases of "true" hermaphroditism (in which babies are born with testicular and ovarian tissue) like Cheryl Chase's, whatever the chromosomes. The determination of an "inadequate" phallus for malehood or of a female reproductive potential seals the fate of these children as constructed girls. Whereas the anatomico-materialist metaphysics of sex in the late nineteenth century made true hermaphrodites an enormous problem for doctors and scientists of that time, the pragmatic, yet anatomically demanding, psychosocial gender-identity theory of today sees "true hermaphrodites" as fairly easily retrofit to either an acceptable male or acceptable female sex/gender.

The logic behind the tendency to assign the female gender in cases of intersexuality rests not only on the belief that boys need "adequate" penises, but also upon the opinion among surgeons that "a functional vagina can be constructed in virtually everyone [while] a functional penis is a much more difficult goal."[30] This is true because much is expected of penises by intersexuality doctors, especially by pediatric urologists. In order for a penis to count as acceptable—"functional"—it must be or have the potential to be big enough to be recognizable as a "real" penis. (Recall, for instance, that Cheryl Chase's penis was considered "monstrously small" as long as she was a boy.) In addition, the "functional" penis is generally expected to have the capability to become erect and flaccid at appropriate times, and to act as the conduit through which urine and semen are expelled, also at appropriate times. The urethral opening is expected to appear at the very tip of the penis. Hypospadias is therefore often treated by constructing a new urethra through to the tip. Typically intersexuality doctors also hope to see penises that are "believably" shaped and colored, and tattoo technology can help with the latter.

As we can see—and as Suzanne Kessler, Morgan Holmes, Cheryl Chase, Ellen Hyun-Ju Lee, Anne Fausto-Sterling, and other critics of present-day treatment protocols have amply documented—intersex surgeons make their decisions and incisions within a heterosexist framework. Lee succinctly states, "Decisions of gender assignment and subsequent surgical reconstruction are inseparable from the heterosexual matrix, which does not allow for other sexual practices or sexualities. Even within heterosexuality, a rich array of sexual practices is reduced to vaginal penetration."[31] Consequently, while much is demanded of a newborn's penis, very little is needed for a surgically constructed vagina to count as "functional." For a constructed vagina to be considered acceptable by intersex surgeons, it basically has to be a hole big enough to fit a typical-sized penis.[32] A constructed "vagina" does not have to have any other abilities. It is not required to be self-lubricating or even to be at all sensitive to count as "functional." (For this reason, Diane Anger's doctors were much more satisfied with her "vagina" than she was.) Nor, to count as functional, does the surgically constructed "vagina" need to change shape the way vaginas often do when women are sexually stimulated. In fact, intersexuality doctors often talk about vaginas in intersex children as the absence of something, as a space, a place to put a penis.[33] That is why opinion holds that "a functional vagina can be constructed in virtually everyone"—because it is relatively easy to surgically construct an insensitive "hole." (Of course, it is not always easy to keep one open; surgical holes often close up with scar tissue.) The decision to "make" a female is therefore considered relatively foolproof, while "the assignment of male sex of rearing is inevitably difficult and should only be undertaken by an experienced team" who can determine if the penis will be adequate for malehood.[34]

The medical treatment of intersexuality today, then, persists in a modernistic vein in several significant and troubling ways. First, unlike postmodernist narratives—such as Anger's, Coventry's, Chase's, and Nicholson's—which question a singular notion of "truth" globally or locally, intersex doctors presume sexual anatomy to be easily measurable as normal or not. They also tell parents of intersexed newborns an excessively simple, straightforward, modernist narrative which claims that the intersex child has a real, true female or male sex and that the doctors only need to discover (rather than invent) which one it is. This storyline is maintained even though doctors know they may find a case of "true hermaphroditism" and may assign a gender which conflicts with the gonadal and chromosomal sex of the child (the kinds of facts parents would tend today to see as marking their child's "real" sex). Suzanne Kessler notes in her ground-breaking analysis of the current treatment of intersexuality: "[T]he message [conveyed to these parents] . . . is that the trouble lies in the doctor's ability to determine the gender, not in the baby's gender per se."[35]

This sort of narrative construction conforms with what Frank calls the restitution narrative, a trope doctors, caretakers, and healthy people strongly prefer to more postmodernistic narratives of chaos or confusion. The restitution narrative says, in its simplest form, "You were healthy; you are sick; but you will be healthy again." In the intersex medical version it reads, "We know what normal sex is; this child has a male or female sex; it just looks confused; we doctors will figure it out and help it not look confused."[36] Not only do intersex doctors tell parents this "real sex" restitution story, they also tell it to themselves; in discourse among themselves they talk about there being a "proper" (right, true) and an "improper" (wrong, false) sex assignment in any given intersex case.[37]

Because of this restitution narrative—in which a sex has *temporarily* blurred—sex borders in intersex medical discourse continue to appear stable in all cases and at all times. A notion of "true sex" is discursively and publicly maintained in intersex cases, as is the one-body-one-sex rule founded so firmly in the nineteenth century. Lee notes that in intersex cases "[p]hysicians present a picture of the 'natural sex,' either male or female, despite their role in actually constructing sex."[38] These medical protocols therefore pay tribute to modernism in their representation of a single "true" female identity—presumed to involve attraction to males, a sizable vagina, and at most a petite phallus—and a single "true" male identity—presumed to involve attraction to females, the absence of the female parts and the presence, most notably, of a sizable, erectile, urine-conducting penis and a scrotum containing two testicles. (Prosthetic testicles are sometimes inserted in a scrotum lacking testes.) Thus modernistic medicine—even when it involves psychosocial theory—in fact remains deeply materialist, reductionist, and determinist in its practical approach to the world, and presumes that a "successful" female gender requires (and is almost guaranteed by) a certain "female" sex anatomy, and that a "successful" male gender requires (and is almost guaranteed by) a certain "male" sex anatomy.

The additional presumption that "restitution" to a "male" or "female" sex is possible and is possible primarily via medical technology is also notably modernistic. Remarkably, the medical-technological approach reigns in intersex medicine despite the fact that intersex experts readily confess that intersexuality is not primarily a medical problem but is a instead a social problem.[39] Intersexual experts presume that, by technologically bringing intersexuals into the anatomical categories of "male" and "female," these individuals will cease to be intersexed—that the social problem will be forestalled or eliminated. As Frank notes, "[M]odernist medicine has regarded suffering as a puzzle to be 'controlled' if not eradicated,"[40] and technology has been the tool of choice for control and eradication of intersex (and other forms of) suffering. The current treatment of intersexuality sees an anatomy-based social "problem" and therefore assumes medical treatment of the anatomy will alleviate that "problem."

The modernistic approach also sees medical technology as inherently good, and any problems arising from the medical treatment—scarring, poor cosmetic results—are presumed to be caused by medical-technological glitches or medical-technological inadequacies that will be eliminated by technological advances.[41] Doubts about the technology never penetrate to question the very use of these medical-technological "fixes." H. Patricia Hynes has noted this kind of phenomenon with regard to agricultural and reproductive biotechnologies: "A mythology encases the technology to make it necessary and acceptable. Once it becomes technically possible, it becomes inevitable." As we have seen above, two more of Hynes's claims could also be applied to intersex surgeries: first, that "those who promote [the technologies] and stand to profit [monetarily] . . . from them, are not those who suffer their risks"; and second, that "the new technology is not presented as one among many solutions to a problem, but as the dominant one. The alternatives to the technology, the 'other road,' are shut out."[42]

Arthur Frank and other medical theorists have noted that doctors tend to think they can and should fix even "social problems" in the only way they know—medically—and intersex doctors are no exception.[43] But not all "problems" that can be subjected to medical treatment should be, particularly when consent is lacking on the part of the potential patient.

This is the "if all you have is a hammer, everything looks like a nail" problem of technology Hynes so aptly describes. Technological approaches tend to shut out discussion of alternative approaches; they quickly appear not only preferable but singular, necessary, and inevitable. Choice in the matter is closed to discussion. Yet, as intersexuals point out, there are choices, and there are also many problems with the surgeries. An inertial faith in technological fixes cannot rule the day.

Finally, the present protocols for the treatment of intersexuality are notably modernistic in their thorough-going paternalism, a phenomenon with a long and now-denigrated history in modern medicine. Paternalism is defined as "interference with a person's autonomy";[44] it assumes that the historically authoritative doctor should have (and should seek) hegemony over patients' knowledge and care. Medical paternalism grew powerfully in the late nineteenth century and continued for much of the twentieth century; doctors assumed that they knew what was best for patients and society and that therefore they should make the primary decisions about a patient's care. In the 1950s, the heyday of paternalism, for example, patients who were diagnosed as having terminal diseases were sometimes not told the truth of their conditions, because their doctors believed that to do so would only depress them and make their last days worse. In the nineteenth-century medical treatment of hermaphroditism, paternalism marked cases like those of Louise-Julia-Anna (in which the patient was told she was a man and told the doctor would refuse to repair her hernia if it meant she would keep having sex with "other" men) and of the Middlesex widow (whose surgeon removed her testicles without ever telling her what he had found).

Usually historians of medicine, medical ethicists, and physicians look back on the days of paternalism and shudder. Yet despite the move of much of medicine out of the ethical quandary of paternalism, in intersex cases the physician still typically plays the role of an almost superhuman benefactor who knows best and who, in his noble efforts to save an otherwise doomed patient, need not tell the patient or parents all he knows. In this respect, the stories of Anger, Coventry, Chase, and Nicholson, full of silences and lies, are still being repeated. Doctors typically make decisions about sex assignment with little genuine discussion with the parents.[45] Parents who will not consent to recommendations are subject to pressure,[46] and even those parents who do agree to the surgeries performed do not realize that they are, by implication, consenting to the doctors' right to choose the sex of their child on the basis of a particular anatomically demanding psychosocial theory of gender identity. Medical journals still publish explicit advice to doctors to withhold facts from intersexed patients and their families. The presumption is that deception and paternalism in these cases is appropriate—even heroic. Thus, in 1995, a medical student was given a cash prize in medical ethics by the Canadian Medical Association for an article specifically advocating deceiving androgen-insensitive patients (including adult patients) about the biological facts of their condition. The author argued that "physicians who withhold information from AIS [androgen insensitivity syndrome] patients are not actually lying; they are only deceiving" because they selectively withhold facts about patients' bodies.[47]

Problems

Intersexual, feminist, and "queer theory" critics of today's dominant treatment protocols point out a number of problems with the modernist medical approach to intersexuality. Most objectionable to many feminist and queer critics is the presumption inherent in these protocols that there is a "right" way to be a male and a "right" way to be a female, and that children who are born challenging these categories should be reconstructed to fit into (and thereby reify) them. These critics ask why homosexuality, "masculinized" girls, or "feminized" boys should be seen as failures of medicine or parenting. Further, they question whether an elite group of physicians should assume the right to decide what counts as the acceptable boundaries of gender identity.

Besides the philosophical-political issue here, there is a more practical one: how does one decide where to put the boundaries on acceptable levels of variation? The definition of "normality" among intersex experts has become so strict that it results in an extraordinary number of risky surgeries on unconsenting children. How many of us would have, as children, fallen into the "abnormal" range given today's criteria among intersex doctors? A skeptical team of physicians in Germany recently took steps to answer that question. That team screened a group of 500 supposedly "normal" men to see how many in fact displayed what is officially deemed hypospadias. (This sample of 500 genitally "normal" men consisted of men who had been "admitted to [the doctors'] hospital between November 1993 and September 1994 for transurethral treatment of benign prostatic hyperplasia and superficial bladder cancer without any previous surgery to the penis.")[48] What the German researchers found is that only 275 (55 percent) of the men could be labeled "normal" according to the medical criteria for penile normality. The remainder of the men in the sample—45 percent—displayed what intersex experts call "hypospadias." Furthermore, "all but 6 patients were not aware of any penile anomaly, all but 1 homosexual patient have fathered children . . . [A]ll patients participated in sexual intercourse without problems and were able to void in a standing position with a single stream." Given this data, the researchers could only conclude that "it remains unclear whether the tip of the glans [penis] truly is the normal site" for the urethral opening.[49] Furthermore, they questioned whether the surgeries performed to "correct" hypospadic penises were necessary, especially given the "significant complication rate" of some of the surgical procedures and given that most of these "hypospadic" men were unaware of and unbothered by their anatomical "problems."[50]

This brings us to another important point: "ambiguous" genitalia do not constitute a metabolic disease. They constitute a failure to fit one particular definition of normality. It is true that whenever a baby is born with "ambiguous" genitalia, doctors need to consider the situation a *potential* medical emergency, because intersexuality may signal one serious metabolic problem, congenital adrenal hyperplasia (CAH). CAH, which can result in "masculinization" of genetic-female fetuses, primarily involves an electrolyte ("salt") imbalance. Treatment of CAH may save a child's life and fertility. At the birth of an intersex child, therefore, adrenogenital syndrome must be quickly diagnosed and treated, or ruled out. Intersex activists take no issue with this. Nonetheless, as medical texts advise, "[o]f all the conditions responsible for ambiguous genitalia, congenital adrenal hyperplasia is the only one that is life-threatening in the newborn period," and even in cases of CAH the "ambiguous" genitalia themselves are not deadly.[51] CAH requires treatment because it is a metabolic disease, but ambiguous genitalia are not. They do not necessarily require medical treatment.

As with CAH's clear medical issue (electrolyte imbalances), doctors now also know, partly from the work of Samuel Pozzi, that the testes of androgen-insensitive people—who look "ambiguous" or "feminized" genitally but are gonadally and chromosomally male—have a relatively high rate of becoming cancerous, and therefore physicians and intersexuals again agree that androgen insensitivity needs to be diagnosed as early as possible so that the testes of androgen-insensitive people can be carefully watched or removed.[52] But the genitalia of an androgen-insensitive person are not diseased. Again, while unusual genitalia may signal a present or potential metabolic threat to health, they in themselves only look different. As we have seen, because of the perception of an intersex birth as a "social emergency," intersex surgeons take license to treat nonstandard genitalia as a medical problem requiring prompt correction. But it is not self-evident that this "problem" should be handed over to pediatric surgeons.

Also problematic in present-day treatment is the infrequency with which the "patient" or his/her family is handed over to psychological counselors or support groups. Despite the effort to make intersexed children look and feel "normal," the way intersexuality is treated by doctors in the United States today inadvertently contributes to many intersexuals' feelings of difference and defectiveness. Such feelings are accentuated because the condition is often shrouded in silence and lies, and because little psychological support is offered to intersexuals and their families. This in turn occurs because of the almost certainly misguided assumption that talking about the reality of intersexuality with the families of intersexuals and with intersexuals themselves will undo all the "positive" effects of the technological efforts aimed at covering up doubts. Yet because of the silences surrounding intersexuality, intersexuals and their families often hear the message that intersexuality is indeed freakish and deeply shameful. B. Diane Kemp—"a social worker with more than 35 years' experience and a woman who has borne androgen insensitivity syndrome (AIS) for 63 years"—noted in her objection to the award-winning article advocating the deception of AIS patients that "secrecy as a method of handling troubling information is primitive, degrading, and often ineffective. Even when a secret is kept, its existence carries an aura of unease that most people can sense . . . Secrets crippled my life."[53]

H. Martin Malin, a professor in clinical sexology and a therapist at the Child and Family Institute in Sacramento, California, has found this theme to be a persistent one in intersexuals' medical experience:

As I listened to [intersexuals'] stories, certain leit motifs began to emerge from the bits of their histories. They or their parents had little, if any, counseling. They thought they were the only ones who felt as they did. Many had asked to meet other patients whose medical histories were similar to their own, but they were stonewalled. They recognized themselves in published case histories, but when they sought medical records, were told they could not be located . . .

The patients I was encountering were not those whose surgeries resulted from life-threatening or seriously debilitating medical conditions. Rather, they had such diagnoses as "micropenis" or "clitoral hypertrophy." These were patients who were told—when they were told anything—that they had vaginoplasties or clitorectomies because of the serious psychological consequences they would have suffered if surgery had not been done. But the surgeries *had* been performed—and they were reporting longstanding psychological distress. They were certain that they would rather have had the "abnormal" genitals they had than the "mutilated" genitals they were given. They were hostile and often vengeful towards the professionals who had been responsible for their care and sometimes, by transference, towards me. They were furious that they had been lied to.[54]

It would appear, from Malin's professional experience, that these surgeries might (in part owing to the lack of follow-up psychological counseling and honesty) be causing the very problems they are supposed to alleviate. Thus, ironically, the typical treatment of intersexuality may well create the feelings it is supposed to prevent.

Imagine the feelings of a typical person born intersexual: highly unusual attention is directed at her/his genitals and general sexual development, but the individual and her/his parents are encouraged to keep the situation a secret and to try not to think about it. Meanwhile the doctor often reveals only some of what she or he knows, while everything possible is done to "normalize" the intersexual's body. Residents and medical students are paraded past the hospitalized intersexual.[55] Would this scenario not necessarily convey to the intersexual that s/he is indeed unacceptable? Diane Marie Anger remembers feeling "like I was the Loch Ness Monster of the human world. I also felt like I was floating in the ocean with 1,000 boats all around me but that no one would throw a line out to me even though I was drowning." Intersexuals commonly express these sorts of feelings. One in Britain writes, for example, "Mine was a dark secret kept from all outside the medical profession (family included), but this [should] not [be] an option because it both increases the feelings of freakishness and reinforces the isolation."[56]

The notion that deception or selective truth-telling will protect the child, the family, or even the adult intersexual is paternalistic as well as naive, of course, and though it may be well intentioned, it goes against the dominant trend in medical ethics as its guidelines are applied to other, similar situations. But lying to a patient about the biology of his or her biological condition is not only ethically questionable, it is bad medicine. This has been noted by Sherri A. Groveman, a woman with AIS. As a young person, Groveman was told she had "twisted ovaries" that had to be removed; in fact, her testes were removed. At the age of twenty, "alone and scared in the stacks of a [medical] library," she discovered the truth of her condition. Then "the pieces finally fit together. But what fell apart was my relationship with both my family and physicians. It was not learning about chromosomes or testes that caused enduring trauma, it was discovering that I had been told lies. I avoided all medical care for the next 18 years. I have severe osteoporosis as a result of a lack of medical attention. This is what lies produce." Groveman pleads with physicians to recognize that "the greatest source of anxiety is not our gonads or karyotype. It is shame and fear resulting from an environment in which our condition is so unacceptable that caretakers lie."[57]

Groveman also argues, as an attorney, "that informed consent laws mandate that the patient know the truth before physicians remove her testes or reconstruct her vagina."[58] It is not at all clear whether all or even most of the intersex surgeries done today involve what would legally and ethically constitute informed consent.[59] Certainly a sizable proportion of the surgeries now being performed on intersexed children would have to, by the rules of medical ethics, be labeled "experimental," because they remain unproven. Indeed, the published literature readily admits to the experimental nature of many of these "cosmetic" surgeries. For example, although these days surgeons reducing clitorises tend to try to operate in such a way as to preserve the sensitivity of the clitoris (not so long ago they only focused on cutting off the "excess"), many of the procedures now done on babies and children are virtually untested in that their success rate is unknown. A team of surgeons from the Children's National Medical Center and George Washington University Medical School in Washington, D.C., reports that in their preferred form of clitoral "recession," "the cosmetic effect is excellent" but "[l]ate studies with assessment of sexual gratification, orgasm, and general psychologic adjustment are unavailable . . . and remain in question."[60] Physicians do not know whether or not this procedure and others like it result in long-term problems such as stenosis (scarring), infections, loss of feeling, or psychological trauma. In fact, "to this date, no studies of clitoral surgery address the long-term results of erotic sexual sensitivity."[61] One team of experts notes that "frustration and uncertainty regarding the long-term outcome for these patients still exist."[62]

Intersexual activists also question whether anyone should have her ability to enjoy sex or physical health risked without her personal consent just because she has a clitoris (or a penis) which statistically falls outside the standard deviation. In other words, even if we did have statistics that showed that particular procedures "worked" a majority of the time—say, for instance, that we had follow-up studies to show that, in 80 percent of all cases in which it was performed, a particular kind of surgical clitoral recession achieved reduction of the clitoris with no loss of sensation or responsiveness and no perception of disfigurement or shame on the part of the patient—even then we would have to face the fact that part of the time it would not work, and we would have to ask whether or not the risk was ever worth assuming on behalf of another person. Certainly there are some procedures (including many intersexuality treatments) that should be described as, at best, experimental and risky to the patient. There are also a significant number for which the decision to proceed should not be left to anyone but the patient him/herself.

Morgan Holmes, who had her "enlarged" clitoris surgically "reduced" when she was six, raises a further question. Even if we did have evidence that these surgeries worked as they are supposed to every time and left a child with unsuspicious-looking, working parts, is it moral for anyone to decide on behalf of a child that a phallus (or vagina) labeled "too big" or "too small" should be surgically altered? Intersexuals pass among themselves the morbid joke that, if the goal is to see to it that an intersex child's genitals matches his/her parents' genitals, why not achieve this by performing "cosmetic" surgery on the parents? After all, they are of legal age to consent to such surgery.

So it is that we see in the intersex medical literature that, while much is expected of the unreconstructed body, far less is demanded of the surgically constructed body, presumably because the latter is seen as inevitably a better alternative to the former. Many of the surgeries designed to make "ambiguous" genitalia look acceptable are very difficult technically and can leave insensitive, painful, disfiguring, or unwieldy scar tissue. The urogenital system is complicated and prone to infection, and like Sven Nicholson, a good number of men who as children had surgeries on their penises (so that they would urinate from the tip of the penis) report lifelong problems with urinary tract infections. These infections can in turn lead to still more scar tissue and more infections. The result can be a penis that not only causes physical pain and illness, but looks extremely "unconvincing."

Assigned-females who have undergone genital surgery are also liable to debilitating and demoralizing urinary and vaginal infections.[63] Surgically constructed vaginas are prone to scarring, and they often have to be artificially dilated and reconstructed to be kept open. In spite of evidence that vaginas constructed after childhood may be no less successful, surgically and psychologically, than those constructed in early childhood, surgeons often perform vaginoplasties on infants and young children because they assume that girl children will not be accepted as girls by their parents if they do not have vaginas. A child too young to perform the repeated "vaginal" dilations generally required after vaginoplasties must have her vagina dilated by a parent or physician. Toddlers do not put up with the dilations well—they see them as extremely invasive—so surgeons rush to perform vaginoplasties on infants.[64] Often by the time puberty is reached, the surgically constructed vagina has closed and requires more surgery.[65] Intersexual activists argue that it makes far more sense, from a surgical and ethical point of view, to wait until a girl is old enough to consent to vaginoplasty and to perform the dilations herself (presuming there is no demonstrable medical reason for vaginoplasty at an earlier age). She might then be advised that, as noted above, even when they do stay open, reconstructed vaginas frequently lack any feeling.

Not surprisingly, some urologists have begun to look at the data and wonder whether at least some of the dominant protocols need revising. For example, does it make sense to do vaginoplasties on infants? At the 1997 meeting of the American Academy of Pediatrics, Dr. David Thomas, a pediatric urologist who practices at St. James's University Hospital and Infirmary in Leeds, England, confessed to his colleagues, "I feel like Daniel stepping into the lion's den. I recognize this may not be a popular message for this audience . . . But I wonder whether we shouldn't be rethinking the philosophy for early vaginal reconstruction."[66] Thomas was concerned about the lack of long-term follow-up studies on the procedures: "So many of these patients are lost to follow-up. If we do this surgery in infancy and childhood, we have an obligation to follow these children up, to assess what we're doing." In any case, Thomas concluded, while "initial surgery for a normal-appearing glans [phallus] and external genitalia is justified," nevertheless "no girl in her childhood needs a functioning vagina."[67] Why take the risk?

In the typical modernist vein, Thomas's opponent at the meeting, Dr. Antoine Khoury, chief of pediatric urology at the Hospital for Sick Children in Toronto, insisted that better techniques would finally make Thomas's concerns moot. He argued that cosmetic genital surgery must be performed on intersex children because "healthy" psychosocial development requires it. Even if this is true, the fact remains that operations done on children before puberty will always be relatively fraught with practical problems.[68] Very small bodies are, by nature, harder to operate on.[69] Pubertal growth spurts sometimes change the look of the previously constructed body in "unhelpful" ways. In part because of these pubertal changes, few people born intersexual who are operated on early in life wind up going through just one surgical procedure. Many find themselves nearly continuous subjects of wanted and unwanted medical attention.

What then of the issue of phallus size? Intersex surgeons are generally unsure whether phalluses left alone will grow or shrink in acceptable ways, because "phallus size at birth has not been reliably correlated with size and function at puberty."[70] This means that "although treatment may not be necessary in the newborn period," intersex experts tend to "ascribe to an early treatment protocol, believing that a phallus of adequate size is important for gender identification and family acceptance."[71] But is it? What of the decision to assign genetic males born with small phalluses as girls, solely because of their phallus size?

As we have seen, virtually all American intersex doctors argue that an XY child born at term with a phallus length less than 2.5 centimeters (when stretched) is better off being made into a girl. But at least two physicians disagree, and have offered evidence from long-term studies in support of their contrary views. These two are Dr. Christopher Woodhouse, a physician at the Institute of Urology and St. George's Hospital in London, and Dr. Justine Reilly Schober, a urologist now at the Hamot Medical Center in Erie, Pennsylvania. In a 1989 article published in the *Journal of Urology* under the title "Small Penis and the Male Sexual Role," Woodhouse and Schober "interviewed and examined 20 patients with the primary diagnosis of micropenis in infancy." Of the postpubertal (adult) subjects, "[a]ll patients were heterosexual and they had erections and orgasms. Eleven patients had ejaculations, 9 were sexually active and reported vaginal penetration, 7 were married or cohabiting and 1 had fathered a child."[72] This flew in the face of the dominant theory, which holds that a boy will not turn out "normally" if he has a very small penis.

From their study, Schober and Woodhouse drew this conclusion: "[A] small penis does not preclude normal male role and a micropenis or microphallus alone should not dictate a female gender reassignment in infancy."[73] More particularly, these doctors found that when parents "were well counseled about diagnosis[,] they reflected an attitude of concern but not anxiety about the problem, and they did not convey anxiety to their children. They were honest and explained problems to the child and encouraged normality in behavior. We believe that this is the attitude that allows these children to approach their peers with confidence." Ultimately, it became clear to Schober and Woodhouse that rather than the child's anatomy being the sine qua non of gender identity and gender crisis, "the strongest influence for all patients was the parental attitude." This in itself does not conflict with the dominant psychosocial theory; Schober and Woodhouse are still saying that the parents' views of the child matter. But rather than telling restitution stories and trying to convince these parents that their children were "really girls," these doctors decided instead to try to convince the parents that their offspring were essentially normal and acceptable. And this worked: "The well informed and open parents . . . produced more confident and better adjusted boys."[74] Being open and supportive—as an alternative to surgeries designed to make these boys into girls—worked quite well. Did the small penises influence these men's sex lives? Yes: "The group was characterized by an experimental attitude to [sexual] positions and methods . . . The group appears to form close and long-lasting relationships. They often attribute partner sexual satisfaction and the stability of their relationships [with women partners] to their need to make extra effort including nonpenetrating techniques."[75]

Schober and Woodhouse are not so naive as to think that all intersex cases will be easy for doctors, patients, or families. But they also do not think that surgery should necessarily be the primary approach to the problem. Schober aptly notes, "Surgery makes parents and doctors more comfortable, but counseling makes people comfortable too, and [it] is not irreversible." Having listened to the pleas of many adult intersexuals, and having seen counseling work well, Schober decided to write a review and critique of intersexual management. In it she asks her colleagues to consider their approaches more carefully: "Simply understanding and performing good surgeries is not sufficient. We must also know when to appropriately perform or withhold surgery. Our ethical duty as surgeons is to do no harm and to serve the best interests of our patient. Sometimes, this means admitting that a 'perfect' solution may not be attainable." This, she notes, may be one of the hardest tasks doctors will face.[76]

History as a Teacher

The roots of the one-body-one-sex rule and medical doctors' role in it were laid down in the late nineteenth century. In spite of all the cultural changes that have occurred, we still have the one-body-one-sex rule, and that rule continues to be driven largely by the same engines that drove it in the nineteenth century: an interest in keeping clear male/female gender distinctions, and a concomitant interest in retaining a clear division between heterosexuality and homosexuality and in supporting what is seen as heterosexuality. Children are sorted early lest "gender identity confusion" lead to "queer" sexuality. Vaginas are specifically "functional" when they can take a "typical" penis, and penises are specifically "functional" when they can be fit into vaginas. The way intersexuals are treated today has much of the same effect intended by the conceptual and practical treatment of the last century: to keep two clear sexes and to retain the notion that heterosexuality is normal and that homosexuality is not.

We know from the history presented in this and other books that the intersex body has never been forced to conform to the one-body-one-sex rule simply for its own sake. Suzanne Kessler sums it up this way: intersexuality does not threaten the patient's life, but rather the patient's culture. Obviously today's medical intersex technologies, particularly surgical and hormonal treatments, are far more elaborate and ambitious than those of the nineteenth century. But the aim of a hundred years ago was much the same, namely, to reduce and ideally to eliminate hermaphroditism using every available tool—conceptual, material, social, rhetorical. The treatment of intersexuality, then as now, draws its motivation in part from concern for the patient. Just as medical men of a hundred years ago worried that people with "masked" ("mistaken") sex would be tormented by the "contradiction" of their "true" and social sexes—and would therefore likely become depressed, frustrated, homosexual, licentious, or even criminal—so, too, do medical professionals today worry about the fate of a child with a "mixed" gender identity. Just as there was a powerful cultural undercurrent to the biomedical treatment of hermaphroditism then, so is there now. The hermaphrodite was and continues to be a person whose body gets caught up in cultural "border wars"—wars over the borders separating males and females, men and women, boys and girls, borders separating the acceptable heterosexual and the disfavored homosexual, borders separating those with authority from those without. "Doubts" inherent in the hermaphrodite's body are still exorcised and shrouded in silence, lest that doubtfulness leak and spread onto the child's identity, the parents and family, medicine and science, and finally the streets. Michel Foucault touched a nerve when he called those of us of the present day "other Victorians." In terms of sex, we have much in common with the Victorians. We still worry a great deal about sex and about order, and about ordering sex. We still insist there be no hermaphrodites, lest they grow and multiply. Cheryl Chase notes, "Intersex is a humanly possible, but (in our culture) socially unthinkable, phenomenon."[77]

The Future

Today's medical, social, and legal systems (in some ways inadvertently, in some ways not) continue to "disappear" hermaphrodites the same way that the conceptual and practical treatments of a hundred years ago aimed to do. This is not just the case with intersexuality, of course. All sorts of diversity are made to disappear, because we assume that it must be traumatic for people to have extra fingers, notable birthmarks, or even big ears. Our compassion tends to be channeled into normalizing. We are not much more comfortable than inhabitants of earlier centuries were with unusual features. Granted, most people would no longer advocate killing babies with relatively unusual anatomies, but we do expect them to endure new forms of suffering to live among us.[78]

No one person created the current treatment of intersexuality, and no one person will change it.[79] But the history of that treatment, and the narratives of those who have received treatment more recently, suggest that it needs changing. All interested parties would benefit from an honest, reflective, and responsible discussion of intersexuality. While we gear up for that larger conversation, consider these additional suggestions for ways in which the treatment of intersexuality in the United States today could be improved.

Certainly any life-threatening metabolic condition associated with intersexuality needs to be attended promptly. Far less nonelective cosmetic surgery, however, can be performed. I do not mean to suggest that surgery should never be offered to an intersexual.[80] It is clear that, today as in the nineteenth century, some intersexuals seek the help of surgeons to have their bodies constructed in particular ways.[81] But medicine at its best empowers rather than overpowers—it provides cosmetic surgery when the patient asks for it and at a point in the patient's life when it is safe and reasonable. Medicine at its best makes available education and choices, and values informed consent. It includes an honest assessment of risks and respect for a patient's choice. Schober reminds her colleagues that "surgery must be based on truthful disclosure and support and permit decision-making by parents and patient."[82]

Further, to conflate "unusual" and "unhealthy" is inaccurate and dangerous. A person might be born with an unusual feature such as violet-colored eyes, or no legs, or hypospadias, but such characteristics do not in themselves make that person unhealthy. Consider, for instance, the experience of M., a man with a hypospadic penis who never chose to have it altered:

> I have never had surgery for it, and am very glad [that I did not] for all the problems I am now aware of. For all the surgical procedures that have been proposed and are still being proposed, it is evident that the perfect procedure has not been found. I can see where scarring even after successful surgery could still be a problem. I have fathered children. The only problem, if it is a problem for it has always been this way, is that I must be careful when I urinate where I aim the urine stream. I always stand. My condition has caused me no problems or embarrassment.

M.'s penis might be unusual—and in truth it is not all that unusual, given how common hypospadias is—but he is by no means unhealthy. Schober notes that there are a number of intersexuals who have stated "that they were glad not to have undergone clitoral reduction surgery. They were not very disturbed by unusual looking genitals and preferred that they had good sensation."[83] Cheryl Chase has asked, "What are genitals for? It is my position that my genitals are for my pleasure."[84]

Some may wonder what will happen if babies born intersexed do not have "corrective" surgery while they are very young. Will they be genderless, or gendered "intersexed"? No. Babies born intersexed can still be assigned the gender of "girl" or "boy" like all other children. As we have seen, it is not our genitals that make our gender identities. Whatever our physical make-up, none of us is fully a "boy" or a "girl" until that identity is made for us by our family and community and embraced by us. One's physical equipment is the signal, not the determinant, of gender identity. Indeed, all gender assignment must ultimately be recognized as preliminary; girls and boys born with the standard parts may grow up to create new and variant gender identities for themselves. So genitals should not be seen as the definitive arbiter of gender identity; in intersexed cases, gender can still be assigned. Schober and Woodhouse showed this with their study of children born with very small penises who were nonetheless raised as boys.

In the case of a child born intersexed, doctors and parents can have an honest conversation about the present and likely future presentation of the child's "sexual" anatomy. Several parameters will hint at the child's future anatomy, including the child's genetic composition, hormone production levels and responsiveness to hormones, and the present anatomical presentation. After an extensive examination of the child's anatomy and an in-depth conversation about sexual anatomy and gender identity, the parents and physicians can then decide which gender assignment makes the most sense. With loving support from parents and doctors, children born intersexed can grow into adults who are comfortable with their bodies, sexually functioning, and able to choose, if they so desire, surgical and hormonal treaments to increase their level of comfort and function.

In addition to delaying consideration of surgery, doctors could also educate themselves about support groups for intersexuals and their families and refer patients to them. Schober asks her colleagues to realize that, in the case of intersexuality as in almost every other "medical" condition, "patients and parents need to be aware of and have access to the societies for persons with their particular problems. Patients feel almost universally relieved to know others with similar problems."[85] With such an approach, perhaps no intersexual would have to lament, as one has, "It's not that my gynecologist told me the truth that angers me (I'd used medical libraries to reach a diagnosis anyway), but that neither I nor my parents were offered any psychological support but were left to flounder in our separate feelings of shame and taboo."[86] In addition to the Intersex Society of North America, many other support groups are forming around the country and the world, including the AIS Support Network in the United Kingdom and Klinefelter Syndrome & Associates in the United States.[87]

All of these groups offer support to intersexuals and their families. Today, many people who find out they were born intersexual are surprised to learn that theirs is not a one-in-a-million experience, and they feel great relief when they locate those who have had similar experiences. We are in an age when doctors, counselors, and others can talk honestly about intersexuality with patients, when we can assure them they are not alone, that they are not freaks, that their conditions are known, understood, and that medicine is there to serve them with education and treatment options.

With such changes in perspective and treatment, the fear of intersexuality may abate. Unusual genitalia do not have to be frightening. As compassion and acceptance for people of varying medical conditions become the norm, the voyeurism and puritanical feelings of shame associated with intersexuality recede. Students may learn about intersexuality in courses on genetics, psychology, and the like, from educators who encourage them to hear and respect the voices of intersexuals. The goal is understanding.

In the last two hundred years, scientists and medical doctors have come to know a tremendous amount about hermaphroditism. They now know much about why some babies arrive in this world with parts that look different from other babies' parts, and they now know of (and employ) a wide variety of techniques designed to change bodies to make them look more "typical." But the accumulation of this knowledge has not taken away the perception that hermaphrodites are strange and troublesome. Four centuries ago, Montaigne declared in his meditations on the fabulous, "If we call prodigies or miracles whatever our reason cannot reach, how many of these appear continually to our eyes! Let us consider through what clouds and how gropingly we are led to the knowledge of most of the things that are in our hands; assuredly we shall find that it is rather familiarity than knowledge that takes away their strangeness."[88] Surely with intersexuals, too, it will be familiarity rather than knowledge that finally takes away their supposed "strangeness."

NOTES

Prologue

1. Dandois, "Un Exemple d'erreur de sexe par suite d'hermaphrodisme appar-ent," *Revue médicale* (Louvain), 5 (1886): 49–52, p. 49. Unless otherwise noted, all translations are my own.

2. Ibid., pp. 50–51.

3. Ibid., p. 49.

4. Ibid.

5. Ibid.

6. For a history of the discovery of "XX" and "XY" chromosomal patterns, see Jane Maienschein, "What Determines Sex? A Study of Converging Approaches, 1880–1916," *Isis,* 75 (1984): 457–480.

7. For a discussion of how our culture has become gene-centered, see Dorothy Nelkin and M. Susan Lindee, *The DNA Mystique: The Gene as a Cultural Icon* (New York: Freeman, 1995).

8. For instance, clitoris size and shape can differ dramatically, or so it seems to many feminist observers. For a fascinating history of varying representations of the clitoris, see Lisa Jean Moore and Adele E. Clarke, "Clitoral Conventions and Trans-gressions: Graphic Representations in Anatomy Texts, c. 1900–1991," *Feminist Stud-ies,* 21 (1995): 255–301.

9. Georges Canguilhem, *The Normal and the Pathological,* trans. Carolyn R. Fawcett (New York: Zone Books, 1991), p. 243.

10. See Denise Grady, "Sex Test," *Discover* (June 1992): 78–82.

11. For a history of how certain hormones came to be thought of as "feminine" or "masculine" hormones, see Nelly Oudshoorn, *Beyond the Natural Body: An Archeol-ogy of Sex Hormones* (New York: Routledge, 1994); note particularly chap. 2. For a

scientist's critique of present-day presumptions about "masculine" and "feminine" hormones, see Anne Fausto-Sterling, *Myths of Gender: Biological Theories about Women and Men,* rev. ed. (New York: Basic Books, 1992), chaps. 4 and 5.

12. I am grateful to Noretta Koertge for reminding me of this.

13. Other authors find similar diversity in other periods; see, for instance, on the Middle Ages, Joan Cadden, *Meanings of Sex Difference in the Middle Ages: Medicine, Science, and Culture* (New York: Cambridge University Press, 1993), and on early modern France, Lorraine Daston and Katharine Park, "The Hermaphrodite and the Orders of Nature: Sexual Ambiguity in Early Modern France," *GLQ (A Journal of Lesbian and Gay Studies),* 1 (1995): 419–438. I suspect that casting a wider net almost invariably results in finding diversity where it has not been found by earlier historians.

1. Doubtful Sex

1. See, for example, Ornella Moscucci, *The Science of Woman: Gynaecology and Gender in England, 1800–1929* (Cambridge: Cambridge University Press, 1990); Elizabeth Fee, "Nineteenth-Century Craniology: The Study of the Female Skull," *Bulletin of the History of Medicine,* 53 (1979): 415–433; Thomas Laqueur, *Making Sex: Body and Gender from the Greeks to Freud* (Cambridge: Harvard University Press, 1990), chap. 5; Robert A. Nye, "Sex Difference and Male Homosexuality in French Medical Discourse, 1830–1930," *Bulletin of the History of Medicine,* 63 (1989): 32–51; Robert A. Nye, *Masculinity and Male Codes of Honor in Modern France* (New York: Oxford University Press, 1993); Cynthia Eagle Russett, *Sexual Science: The Victorian Construction of Womanhood* (Cambridge: Harvard University Press, 1989).

2. See E. Goujon, "Étude d'un cas d'hermaphrodisme bisexuel imparfait chez l'homme," *Journal de l'anatomie et de la physiologie normales et pathologiques de l'homme et des animaux,* 6 (1869): 599–616. For an English translation of parts of this article, see Herculine Barbin, *Herculine Barbin: Being the Recently Discovered Memoirs of a Nineteenth-Century French Hermaphrodite,* intro. Michel Foucault, trans. Richard McDougall (New York: Pantheon, 1980), pp. 128–144.

3. Goujon, "Étude," p. 606.

4. Ibid., p. 599.

5. Ibid., pp. 607–608.

6. Ibid., p. 600.

7. Some authors have been confused about the interval that existed between the start of the memoirs and the suicide. Barbin's memoirs begin, "I am twenty-five years old, and, although I am still young, I am beyond any doubt approaching the hour of my death" (Barbin, *Herculine Barbin,* p. 3). Since Barbin was born in November 1838, we can assume the memoirs began in late 1863 or in 1864. It is not clear when Barbin ceased writing. Probably because Barbin did not commit suicide until four years after the memoirs start, several commentators on the case confused the timing of the change of civil status (June 1860) and the suicide (February 1868). Lutaud, for

instance, reported that the suicide occurred three years after the change of civil status when in fact it occurred seven and a half years later. (See August Lutaud, "De L'Hermaphrodisme au point de vue médico-légal. Nouvelle observation. Henriette Williams," *Journal de médecine de Paris* [1885]: 386–396, p. 388.)

8. See Ambroise Tardieu, *Question médico-légale de l'identité dans ses rapports avec les vices de conformation des organes sexuels,* 2d ed. (Paris: J. B. Baillière et Fils, 1874).

9. Barbin, *Herculine Barbin,* p. 103.

10. Ibid., p. 20; original emphasis.

11. Ibid., p. 56; original emphasis.

12. Ibid., p. 52; original emphasis.

13. Ibid., pp. 150–151.

14. Ibid., p. 89.

15. Ibid., p. 104.

16. See, for instance, the commentaries on Barbin's case in Tardieu, *Question médico-légale de l'identité;* Lutaud, "De L'Hermaphrodisme," p. 388; Gérin-Roze, "Un Cas d'hermaphrodisme faux," *Bulletins et mémoires de la société médicale des hôpitaux,* 3d ser., 1 (1884): 369–373, p. 372; Dandois, "Un Exemple d'erreur de sexe par suite d'hermaphrodisme apparent," *Revue médicale,* 5 (1886): 49–52, discussion on p. 52; Brouardel, "Hermaphrodisme; impuissance; type infantile," *Gazette des hôpitaux civils et militaires,* 60 (1887): 57–59, p. 57; Jean Tapie, "Un Cas d'erreur sur le sexe. Malformation des organes génitaux externes; pathogénie de ces vices de conformation," *Revue médicale de Toulouse,* 22 (1888): 301–313, p. 306; Samuel Pozzi, "De L'Hermaphrodisme," *Gazette hebdomadaire de médecine et de chirurgie,* 2d ser., 27, no. 30 (1890): 351–355, p. 354, n. 2; Louis Blanc, *Les Anomalies chez l'homme et les mammifères* (Paris: J.-B. Baillière et Fils, 1893), pp. 195–196; Bacaloglu and Fossard, "Deux cas de pseudo-hermaphroditisme (gynandroïdes)," *Presse médicale,* 2, no. 97 (1899): 331–333, p. 331; Xavier Delore, "Des Étapes de l'hermaphrodisme," *L'Écho médicale de Lyon,* 4, no. 7 (1899): 193–205 and no. 8 (1899): 225–232, p. 230; Franz Neugebauer, "Cinquante cas de mariages conclus entre des personnes du même sexe, avec plusieurs procès de divorce par suite d'erreur de sexe," *Revue de gynécologie et de chirurgie abdominale,* 3 (1899): 195–210, p. 195; Franz Neugebauer, *Hermaphroditismus beim Menschen* (Leipzig: Werner Klinkhardt, 1908).

17. Neugebauer's *Hermaphroditismus beim Menschen* purports to include a reproduction of a portrait of Barbin, but as Michel Foucault has noted, "the printer by mistake put Alexina's name under a portrait that is obviously not her own" (Foucault in Barbin, *Herculine Barbin,* p. xiv, n. 1). I suspect this error may have arisen from the juxtaposition of a vague reference to Alexina and that same picture in an earlier text (Émile Laurent, *Les Bisexués. Gynécomastes et hermaphrodites* [Paris: Georges Carré, 1894]) on which Neugebauer relied.

18. Fancourt Barnes, [report of a living specimen of a hermaphrodite, meeting of April 25, 1888], *British Gynaecological Journal* 4 (1888): 205–212, p. 205.

19. Ibid., p. 212.

20. Ibid., p. 206.

21. Ibid., p. 208.

22. Ibid.

23. Ibid.

24. Ibid., p. 210.

25. Ibid.

26. Ibid., p. 212.

27. See Roy Porter, "The Patient's View: Doing Medical History from Below," *Theory and Society,* 14 (1985): 175–198.

28. For a review of how the preponderance of men in science and medicine influenced visions of sex, gender, and sexuality from the eighteenth to the twentieth centuries, see Ludmilla Jordanova, *Sexual Visions: Images of Gender in Science and Medicine between the Eighteenth and Twentieth Centuries* (Madison: University of Wisconsin Press, 1989).

29. For discussions of women's attempts to be recognized as medical professionals, see Thomas Neville Bonner, *Becoming a Physician: Medical Education in Britain, France, Germany, and the United States, 1750–1945* (New York: Oxford University Press, 1995), pp. 207–213, 312–315; see also W. F. Bynum, *Science and the Practice of Medicine in the Nineteenth Century* (Cambridge: Cambridge University Press, 1994), pp. 206–208, and Moscucci, *Science of Woman,* pp. 71–73.

30. Moscucci, *Science of Woman,* p. 73. For a discussion of the emergence of these medical specialties in Paris, see George Weisz, "The Development of Medical Specialization in Nineteenth-Century Paris," in Ann LaBerge and Mordechai Feingold, eds., *French Medical Culture in the Nineteenth Century* (Atlanta: Rodopi, 1994), pp. 149–188.

31. Ian Hacking traces a similar amplification cycle in this century in the history of multiple personality disorder, although his interpretations of that phenomenon differ somewhat from my interpretation of the history of hermaphroditism. See Hacking's *Rewriting the Soul: Multiple Personality and the Sciences of Memory* (Princeton: Princeton University Press, 1995).

32. It also seems that greater publicity about hermaphroditism may have led some hermaphrodites to realize their sex might be "mistaken" and to seek out medical consultation. See, for instance, the case of Joseph-Marie X. in Chapters 3 and 4. The prolific Dr. Pierre Garnier was specifically sought out by a young, self-educated "mistaken male" who diagnosed his own condition after reading Garnier's popular books on the subject. Garnier told the story this way: "Declared, registered, and raised a girl, he [the subject of "mistaken" sex] lost his mother in good time and thus it was not possible for it [the "mistake"] to be cleared up by her in his state of puberty. Since that time, he has shown a lively sentiment of love for his [female] companions, and, in the absence of menstruation, he conceived some doubts about his true sex, even more so as his penis became more and more voluminous and

prominent. Spontaneous erections, nocturnal pollutions confirmed his doubts, but nonetheless he was always obliged [by his anatomical conformation] to crouch in order to urinate so as not to soil himself, and a crevice existed under the penis as in his feminine companions. Living only with his father in a small town, without [other] family, he thus remained undecided for a long time, without any illness which would have furnished him an occasion to confide in a medical man. It was in reading my various works on generation and on hermaphrodism in particular that he recognized his true sex and the hypospadias which afflicted him." Pierre Garnier, "Du Pseudo-hermaphrodisme comme impédiment médico-légale à la déclaration du sexe dans l'acte de naissance," *Annales d'hygiene publique et de médecine légale*, 3d ser., 14 (1885): 285–293, p. 290.

33. Russett, *Sexual Science*, p. 7.

34. Patrick Geddes and J. Arthur Thomson, *The Evolution of Sex* (London: Walter Scott, 1889), p. 267.

35. Garnier, "Du Pseudo-hermaphrodisme," p. 287.

36. Jonathan Hutchinson, "A Note on Hermaphrodites," *Archives of Surgery*, 7 (1896): 64–66, p. 64.

37. Russett, *Sexual Science*, p. 48.

38. Judith Butler, *Gender Trouble: Feminism and the Subversion of Identity* (New York: Routledge, 1990), p. 110.

39. I use the expression "gonad" to stand for ovaries, testes, and ovotestes because it simplifies my narrative, especially when the nature of the sex gland had yet to be determined in a given case. I prefer not to use the expression "sex gland," because it connotes an endocrinological understanding of the gonad, and, as Chapter 5 documents, until the twentieth century, in matters of hermaphroditism gonads were of interest only in terms of their anatomy and not in terms of their physiology.

40. On the "scientizing" of medicine in the nineteenth century, particularly in Germany, see Bynum, *Science and the Practice of Medicine*, chaps. 4 and 5; see also Bonner, *Becoming a Physician*, chap. 10, and Chapter 5 below. On the notion that men and women differed because of evolutionary roles, see Russett, *Sexual Science*, pp. 11–12 and chap. 5.

41. Th. Tuffier and A. Lapointe, "L'Hermaphrodisme. Ses variétés et ses conséquences pour la pratique médicale (d'après un cas personnel)," *Revue de gynécologie et de chirurgie abdominale*, 17 (1911): 209–268, p. 256.

42. See Richard Goldschmidt, "Intersexuality and the Endocrine Aspect of Sex," *Endocrinology* (Philadelphia), 1 (1917): 433–456. Goldschmidt notes that he "proposed the use of the terms intersexe [*sic*], intersexual, [and] intersexuality" as a replacement for the term "sex-integrades because the former terms can be used in all scientific languages, whereas the latter must be translated, e.g., 'Sexuelle Zwischenstufen' in German"; see p. 437n. For additional discussion of the term and its origin, see also Richard Goldschmidt, *The Mechanism and Physiology of Sex Determination*, trans. William J. Dakin (London: Methuen & Co., Ltd., 1923), p. 77; and Hugh

Hampton Young, *Genital Abnormalities, Hermaphroditism, and Related Adrenal Diseases* (Baltimore: Williams and Wilkins Company, 1937), p. 50.

43. See Goldschmidt, *The Mechanism and Physiology of Sex Determination,* p. 249. For an early use of the term "intersex" to refer to homosexuality, see Xavier Mayne, pseudonym for Edward Irenaeus Prime Stevenson, *The Intersexes. A History of Similisexualism as a Problem in Social Life* (privately printed, n.d., preface dated 1908; reprint, New York: Arno Press, 1975).

44. The quotation is from Joseph H. Keifer, "The Hermaphrodite as Depicted in Art and Medical Illustration," *Urological Survey,* 17 (April 1967): 65–70, p. 65. It is not uncommon for medical texts like Keifer's to recount this myth before discussing the scientific and medical aspects of hermaphroditism.

45. Ovid, *Metamorphoses,* books 1–4, trans. D. E. Hill (Oak Park, Ill.: Bolchazy-Carducci, 1985), book 4, lines 371–372.

46. Ibid., lines 374–376.

47. See Joan Cadden, *Meanings of Sex Difference in the Middle Ages: Medicine, Science, and Culture* (New York: Cambridge University Press, 1993); see also Lorraine Daston and Katharine Park, "The Hermaphrodite and the Orders of Nature: Sexual Ambiguity in Early Modern France," *GLQ (A Journal of Lesbian and Gay Studies),* 1 (1995): 419–438, esp. pp. 420–423.

48. On this, see Cary J. Nederman and Jacqui True, "The Third Sex: The Idea of the Hermaphrodite in Twelfth-Century Europe," *Journal of the History of Sexuality,* 6 (1996): 497–517, pp. 506–507. For a discussion of John Donne's use of hermaphrodite imagery and the idea of an androgynous deity, see Frances M. Malpezzi, "Adam, Christ, and Mr. Tilman: God's Blest Hermaphrodites," in *American Benedictine Review,* 40 (1989): 250–260. Malpezzi notes Donne's references to the verse of Genesis (1:27) that led some early Christian theologians to believe that God had first created humans in hermaphroditic form: "So God created man in his own image, in the image of God created he him; male and female created he them" (from Malpezzi, p. 252).

49. For an overview of hermaphrodite history, see Julia Epstein, "Either/Or—Neither/Both: Sexual Ambiguity and the Ideology of Gender," *Genders,* 7 (1990): 99–142; note, however, that Epstein's work is perhaps too ambitious in its scope and needs to be read with other interpretations and closer studies of the various periods.

50. Nederman and True ("The Idea of the Hermaphrodite," pp. 501–502) claim that "there is much evidence to suggest that hermaphrodites were tolerated (at least so long as they conformed to one or another gender role)" in the twelfth century, but unfortunately they do not offer references to that evidence.

51. Daston and Park, "The Hermaphrodite," p. 419.

52. Ibid., p. 429.

53. See Isidore Geoffroy Saint-Hilaire, *Histoire générale et particulière des anomalies de l'organization chez l'homme et les animaux . . . ou Traité de tératologie* (Paris: J.-B. Baillière, 1832–36). For a summary review of this text, see Alice Domurat Dreger,

"'Nature Is One Whole': Isidore Geoffroy Saint-Hilaire's *Traité de tératologie*" (M.A. thesis, Indiana University, 1993).

54. J. W. Ballantyne, "Hermaphroditism," in Chalmers Watson, ed., *Encyclopaedia Medica,* vol. 4 (Edinburgh: William Green and Sons, 1900), p. 491. Isidore Geoffoy Saint-Hilaire and his contemporary teratologists agreed with this principle. See also Tapie, "Un Cas d'erreur," pp. 302, 307–313; and Bacaloglu and Fossard, "Deux cas de pseudo-hermaphroditisme," p. 333.

55. See, for example, Oscar Hertwig, *Text-Book of the Embryology of Man and Mammals,* trans. from the third edition by Edward L. Mark (London: Swan Sonnenschein & Company, 1892), p. 374: "It is only in the very early stages that it is impossible to distinguish whether the germinal epithelium will be developed into testis or ovary. Differences soon appear, which allow a positive determination."

56. On this, see Delore, "Des Étapes," pp. 201ff.; Xavier Delore, "De L'Hermaphrodisme dans l'histoire ancienne et dans la chirurgie moderne," *Journal des sciences médicales de Lille,* 2 (1899): 63–70, p. 67; Hertwig, *Text-Book of the Embryology,* pp. 353–411; and Charles Sedgwick Minot, *Human Embryology* (New York: W. Wood, 1892), pp. 490–492.

57. See pp. 147–153 in Nancy Tuana, "The Weaker Seed: The Sexist Bias of Reproductive Theory," in Nancy Tuana, ed., *Feminism and Science* (Bloomington: Indiana University Press, 1989), pp. 147–171. See also Marie-Hélène Huet, *Monstrous Imagination* (Cambridge: Harvard University Press, 1993), pp. 3–4.

58. See, for example, Tapie, "Un Cas d'erreur," p. 309: "the appearance of every embryo is rather feminine." See also Arène, "Un Cas de pseudo-hermaphrodisme," *Loire médical,* 14 (1895): 187–195, esp. p. 192: "the external genital organs have kept an embryonic disposition, and consequently [look] feminine."

59. See, for example, Minot, *Human Embryology,* pp. 517–518. For a critique of the common way of thinking about male and female development, see Anne Fausto-Sterling, *Myths of Gender: Biological Theories about Women and Men,* rev. ed. (New York: Basic Books, 1992), pp. 77–85.

60. Minot, *Human Embryology*, p. 520.

61. This challenges Laqueur's contention that by the nineteenth-century a "two-sex" model, according to which men and women were understood to be fundamentally different, had taken root in Western science, medicine, and culture, completely displacing an older "one-sex" model. (See Laqueur, *Making Sex.*) I do, however, agree wholeheartedly with Laqueur's major point that sex—anatomy and physiology—gets "constructed" just as gender does.

62. Rosemarie Garland Thomson, "Introduction: From Wonder to Error—A Genealogy of Freak Discourse in Modernity," in Rosemarie Garland Thomson, ed., *Freakery: Cultural Spectacles of the Extraordinary Body* (New York: New York University Press, 1996), pp. 1–19, p. 4. For an account of the domestication of the monstrous in evolutionary theories, see Evelleen Richards, "A Political Anatomy of Monsters, Hopeful and Otherwise: Teratogeny, Transcendentalism, and Evolutionary Theoriz-

ing," *Isis,* 85 (1994): 377–411. For a discussion of Diderot's domestication of the hermaphrodite in the *Encyclopédie,* see James McGuire, "La Représentation du corps hermaphrodite dans les planches de l'*Encyclopédie,*" *Recherches sur Diderot et sur l'Encyclopédie,* 11 (October 1991): 109–129.

63. See Elizabeth Grosz, "Intolerable Ambiguity: Freaks as/at the Limit," in Thomson, ed., *Freakery,* pp. 55–66; p. 58.

64. Geoffroy thought, for instance, that if teratological experiments showed physical trauma to be a cause of serious monstrosities in chick embryos, then that knowledge might be used to prevent monstrosities or at least be used to instigate searches for violent culprits (like wife-beaters) when a woman gave birth to a monstrous child. See Isidore Geoffroy Saint-Hilaire, *Histoire générale et particulière des anomalies,* vol. 3, pp. 582–583, and Dreger, "'Nature Is One Whole,'" pp. 156–157.

65. This is discussed in more detail in the Epilogue; also see Epilogue for additional references to present-day medical literature on intersex states.

66. Keith L. Moore, *Before We Are Born: Basic Embryology and Birth Defects,* 3d ed. (Philadelphia: W. B. Saunders Company, 1989), p. 195. See also Ethel Sloane, *Biology of Women,* 3d ed. (Albany, N.Y.: Delmar Publishers, 1993), p. 168.

67. For a discussion of the genetic aspects of true hermaphroditism, see Margaret W. Thompson, Roderick R. McInnes, and Huntington F. Willard, *Thompson & Thompson Genetics in Medicine,* 5th ed. (Philadelphia: W. B. Saunders Company, 1991), pp. 242–243.

68. Moore, *Before We Are Born,* p. 195.

69. On female pseudohermaphroditism, see Thompson, McInnes, and Willard, *Thompson & Thompson Genetics in Medicine,* p. 244; Sloane, *Biology of Women,* pp. 169–170; Moore, *Before We Are Born,* p. 195; "Intersex States," in Robert Berkow, editor-in-chief, *The Merck Manual of Diagnosis and Therapy,* 15th ed. (Rahway, N.J.: Merck Sharp and Dohme Research Laboratories, 1987), pp. 1962–63; Thomas E. Andreoli et al., *Cecil Essentials of Medicine,* 3d ed. (Philadelphia: W. B. Saunders Company, 1993), p. 494.

70. Experts today believe that androgen insensitivity can range from partial to complete. My description here is of the more complete form of androgen insensitivity.

71. The phrase "key androgen receptor" is from Berkow, "Intersex States," p. 1962. On AIS, see also Moore, *Before We Are Born,* p. 195; Thompson, McInnes, and Willard, *Thompson & Thompson Genetics in Medicine,* pp. 243–244; Sloane, *Biology of Woman,* pp. 168–169; and James E. Griffin, "Androgen Resistance: The Clinical and Molecular Spectrum," *New England Journal of Medicine,* 326 (February 27, 1992): 611–618.

72. On 5-AR deficiency, see Griffin, "Androgen Resistance"; see also Berkow, "Intersex States," p. 1962. On 5-AR deficiency and AIS, see also Jean D. Wilson, "Syndromes of Androgen Resistance," *Biology of Reproduction,* 46 (1992): 168–173.

73. A discussion of Klinefelter's is provided in most medical reviews of intersex conditions. For a history of the discovery of the syndrome by Harry F. Klinefelter

himself, see "Klinefelter's Syndrome: Historical Background and Development," *Southern Medical Journal,* 79 (1986): 1089–93; see also Klinefelter, "Background of the Recognition of Klinefelter's Syndrome as a Distinct Pathologic Entity," *American Journal of Obstetrics and Gynecology,* 116 (1973): 436–437.

74. On hypospadias, see Moore, *Before We Are Born,* pp. 195–198; Berkow, "Intersex States," p. 1962. For references to articles of historical interest with specific regard to hypospadias, see Alice Domurat Dreger, "Doubtful Sex: Cases and Concepts of Hermaphroditism in France and Britain, 1868–1915" (Ph.D. diss., Indiana University, 1995), p. 44, n. 5.

75. For these statistics, see respectively Andreoli et al., *Cecil Essentials,* p. 494; Sloane, *Biology of Woman,* p. 169; and Thompson, McInnes, and Willard, *Thompson & Thompson Genetics in Medicine,* p. 244.

76. See Gilbert Herdt, "Mistaken Sex: Culture, Biology and the Third Sex in New Guinea," in Gilbert Herdt, ed., *Third Sex, Third Gender: Beyond Sexual Dimorphism in Culture and History* (New York: Zone Books, 1996), pp. 419–445.

77. Sloane, *Biology of Woman,* p. 168. According to Denise Grady, a study of over 6,500 women athletes competing in seven different international sports competitions showed an incidence of intersexuality of one in 500 women, but unfortunately Grady does not provide a reference to the published data of that study. (See Denise Grady, "Sex Test," *Discover* [June 1992]: 78–82.) That sampled population should not be taken as representative of the whole population, but this number is certainly higher than most people would expect.

78. Anne Fausto-Sterling, pers. comm. and "Building Bodies: Biology and the Social Construction of Sexuality" (forthcoming), chap. 2.

79. The highest modern-day estimate for frequency of sexually ambiguous births comes from the Johns Hopkins psychologist John Money, who has posited that as many as 4 percent of live births today are of "intersexed" individuals (cited in Anne Fausto-Sterling, "The Five Sexes," *The Sciences,* 33 [March/April 1993]: 20–25, p. 21). Money's categories tend to be exceptionally broad and poorly defined, and not representative of what most medical experts today would consider to be "intersexuality."

80. These estimates for the incidence of cystic fibrosis and Down syndrome are taken from Thompson, McInnes, and Willard, *Thompson & Thompson Genetics;* for cystic fibrosis, see p. 67; for Down syndrome, see p. 224.

81. Thomson, "Introduction: From Wonder to Error," p. 1.

82. Mary Douglas, *Purity and Danger: An Analysis of Concepts of Pollution and Taboo* (London: Ark, [1966] 1984); p. 115. Since about the mid-1980s, historians of medicine and other humanistic scholars have been fascinated by the history of "the body" (and various kinds of bodies and various parts of the body). For an introduction to the literature, see Roy Porter, "History of the Body," in Peter Burke, ed., *New Perspectives on Historical Writing* (University Park, Pa.: Pennsylvania State University Press, 1992), pp. 206–232. The Zone series, *Fragments for a History of the Human Body,* also makes a good starting point. The collection edited by Jennifer Terry and

Jacqueline Urla provides a sense of how different approaches to bodies can be; see *Deviant Bodies: Critical Perspectives on Difference in Science and Popular Culture* (Bloomington: Indiana University Press, 1995). For an analysis of how language works to "construct" bodies and for a review of various ways to think about bodies in the history of medicine, see David Michael Levin and George F. Solomon, "The Discursive Formation of the Body in the History of Medicine," *Journal of Medicine and Philosophy,* 15 (1990): 515–537. For a sweeping review of the major body-metaphors in the history of Western culture, see Anthony Synnott, "Tomb, Temple, Machine, and Self: The Social Construction of the Body," *British Journal of Sociology,* 43 (1992): 79–110. In keeping with Simone de Beauvoir's observation that "one is not born a woman, one becomes one," feminists have long been interested in documenting the construction of women's bodies; see, for example, Russett, *Sexual Science;* Moscucci, *Science of Woman,* pp. 102–108; Londa Schiebinger, "Skeletons in the Closet: The First Illustrations of the Female Skeleton in Eighteenth-Century Anatomy," *Representations,* 14 (1986): 42–81; see also the collection edited by Mary Jacobus, Evelyn Fox Keller, and Sally Shuttleworth, *Body/Politics: Woman and the Discourses of Science* (New York: Routledge, 1990). Historians have also traced the history of racial concepts as they have been constructed through bodily representations; see, for example, chapters 4 and 5 in Londa Schiebinger, *Nature's Body: Gender in the Making of Modern Science* (Boston: Beacon Press, 1993); Stephen Jay Gould, *The Mismeasure of Man* (New York: W. W. Norton, 1981); see also Sander Gilman, "The Jewish Body: A 'Footnote,'" *Bulletin of the History of Medicine,* 64 (1990): 588–602. Gilman has done a particularly good job of showing the connections between bodies in art, bodies in medicine, and bodies in literature; see, for example, his "Black Bodies, White Bodies: Toward an Iconography of Female Sexuality in Late Nineteenth-Century Art, Medicine, and Literature," *Critical Inquiry* 12 (1985): 204–242; see also his *Sexuality: An Illustrated History* (New York: John Wiley & Sons, 1989). For an example of how a history of the representations of a single body part can be traced in detail, see Lisa Jean Moore and Adele E. Clarke, "Clitoral Conventions and Transgressions: Graphic Representations in Anatomy Texts, c. 1900–1991," *Feminist Studies,* 21 (1995): 255–301.

83. Delore, "Des Étapes," pp. 231–232.

84. It has not escaped my attention that these problems of objectivity and representation inhere in my own work. I subscribe to the working model of "objectivity as responsibility" presented in Lisa M. Heldke and Stephen H. Kellert, "Objectivity as Responsibility," *Metaphilosophy,* 26 (1995): 360–378.

2. Doubtful Status

1. For instance, a previously examined hermaphroditic "woman" was scheduled to be shown to Rudolf Virchow, but she declined to return for more examinations; see Franz Neugebauer, "Hermaphrodism in the Daily Practice of Medicine; Being Infor-

mation upon Hermaphrodism Indispensable to the Practitioner," *British Gynaecological Journal,* 29 (1903): 226–263, p. 235. For other examples, see Chapter 4.

2. For examples of instances in which the question of complicity arose, see the case histories of Louise-Julia-Anna and Madame X. of Angers in Chapter 4.

3. Barbin's case illustrates the way in which "hermaphrodite" was a changeable identity rather than an ontological category: if Barbin had been recognized at birth to be a hypospadic male, he would never have qualified as a "hermaphrodite," but merely as a "male hypospade."

4. Similarly, Marie-Madeleine Lefort, available in person in life and after death, garnered much attention. Lefort's case is described below. On Barbin, see references in Chapter 1.

5. Those interested in the scholarly pursuit jokingly known among historians of medicine as "retrospectroscopy"—that is, diagnosing long-dead (usually famous) people with current-day medical classifications—generally will diagnose Barbin as having been 5-AR deficient. (I take the term "retrospectroscopy" from the discussion in W. F. Bynum, *Science and the Practice of Medicine in the Nineteenth Century* [New York: Cambridge University Press, 1994], p. 208.) Retrospectroscopic accounts of hermaphroditism by present-day physicians may be found in Maria I. New and Elizabeth S. Kitzinger, "Pope Joan: A Recognizable Syndrome," *Journal of Clinical Endocrinology and Metabolism,* 76 (1993): 3–13; J. M. Walshe, "Tutankhamun: Klinefelter's or Wilson's?" *Lancet,* 1 (1973): 109–110; R. Bakan, "Queen Elizabeth I: A Case of Testicular Feminization?" *Medical Hypotheses* 17 (1985): 277–284; Peter J. Capizzi and Charles E. Horton, "A Case of Colonial Gender Conflict: Thomas (Thomasine) Hall," *Annals of Plastic Surgery,* 23 (1989): 320–322; and Christos S. Bartsocas, "Goiters, Dwarfs, Giants, and Hermaphrodites," *Progress in Clinical and Biological Research,* 200 (1985): 1–18.

6. P. D. Handyside, "Account of a Case of Hermaphrodism," *Edinburgh Medical and Surgical Journal,* 43 (1835): 313–318, p. 314.

7. Ibid., p. 318.

8. See "Certificates of a Very Rare Specimen of Hermaphroditism" (n.d.), pamphlet, British Library catalog number Cup.366.e.20.

9. See Robert Bogdan, *Freak Show: Presenting Human Oddities for Amusement and Profit* (Chicago: University of Chicago Press, 1988).

10. See Irving Wallace and Amy Wallace, *The Two* (New York: Simon and Schuster, 1978); Alice Domurat Dreger, "The Limits of Individuality: Ritual and Sacrifice in the Lives and Medical Treatment of Conjoined Twins," *Studies in History and Philosophy of Science,* forthcoming.

11. On Lefort, see Trélat, "De L'Hermaphrodisme féminin (leçon recueillie par M. G. Marchant)," *Journal des connaissances médicales pratiques et de pharmacologie,* 3d ser., 2 (1880): 41–42, 58–59, p. 59; Albert Leblond, "Du Pseudo-hermaphrodisme comme impédiment médico-légal a la déclaration du sexe dans l'acte de naissance," *Annales d'hygiene publique et de médecine légale,* 3d ser., 14 (1885): 293–302, p. 299;

François Jules Octave Guermonprez, "Une Erreur de sexe, avec ses conséquences," *Journal des sciences médicales de Lille* 2 (1892): 337–349, 361–376, p. 362, n. 1; Louis Blanc, *Les Anomalies chez l'homme et les mammifères* (Paris: J.-B. Baillière et Fils, 1893), pp. 200–201; Arène, "Un Cas de pseudo-hermaphrodisme," *Loire médical,* 14 (1895): 187–195, p. 194; and Houel, "Pièces d'hermaphrodites conservées au musée Dupuytren," *Bulletin de la société d'anthropologie de Paris,* 3d ser., 4 (1881): 554–558, pp. 554–555.

12. James Young Simpson, "Hermaphroditism, or Hermaphrodism," in Robert Bentley Todd, ed., *The Cyclopaedia of Anatomy and Physiology,* vol. 2 (London: Sherwood, Gilbert and Piper, 1839), pp. 684–738, esp. p. 703. (This article was reprinted in Sir W. G. Simpson, ed., *The Works of Sir James Y. Simpson, Bart.,* vol. 2, *Anaesthesia, Hospitalism, Hermaphroditism, and a Proposal to Stamp Out Small-Pox and Other Contagious Diseases* [New York: D. Appleton & Co., 1872].)

13. See A. Campbell Geddes, "A Specimen Illustrating Pseudo-Hermaphroditism," *British Medical Journal,* 2 (September 28, 1912): 769–770, p. 769. Note also the case discussed on page 87 of Alcide Benoist, "Rapport sur un cas d'hermaphrodisme," *Annales d'hygiene publique et de médecine légale,* 3d ser., 16 (1886): 84–87. There a hermaphrodite seeks to be declared male in part "to be able to meet the material needs of his existence."

14. A. Lutaud, "De L'Hermaphrodisme au point de vue médico-légal. Nouvelle observation. Henriette Williams," *Journal de médecine de Paris* (1885): 386–396, p. 395. Lutaud also reported that Williams "was very happy to learn that surgery would not be completely impotent [in the matter] of giving her genital organs a more masculine appearance."

15. J. Halliday Croom, "Two Cases of Mistaken Sex in Adult Life," *Transactions of the Edinburgh Obstetrical Society,* 24 (1898–1899): 102–104, p. 103. This account in turn is based on the narrative found in observation 2 of Franz Neugebauer, "Cinquante cas de mariages conclus entre des personnes du même sexe, avec plusieurs procès de divorce par suite d'erreur de sexe," *Revue de gynécologie et de chirurgie abdominale,* 3 (1899): 195–210. Neugebauer in turn credits the story to Badaloni.

16. Ernest Hart, ed., [Exhibit by Mr. Brian Rigden of Spurious Hermaphroditism], *British Medical Journal,* 2 (December 2, 1882): 1119.

17. For instance, in his 1903 report on hermaphroditism to the British Gynaecological Society, Franz Neugebauer included "as curiosities . . . the following exceptional cases from among those I have collected." It read: "There have been two instances in which a soldier has proved to be a woman: in one, menstruation betrayed the fact, and the individual had to leave the army and adopt female attire; the other gave birth to a child; a similar case gave rise to the well-known phrase: 'Mas mulier, monachus, mundi mirabile monstrum [manly woman, monk, the world's marvelous wonder].' Again, a youth, servant in a monastery, was, for having stolen a silver cup, condemned to be whipped, naked, before the whole community, but begged not to be stripped and put to shame as he knew he was a girl; this was the case, and the true

sex and existence of menstruation having been ascertained, not only was the punishment remitted, but the young person was provided with female clothes and a dowry, and her true sex having, thanks to the accident of her theft, been recognised, she was married to a wine seller, by whom she had several children. Morache, in his recent work, 'Le mariage,' relates that for several years he was at college with a fellow pupil, an excellent comrade, extremely intelligent, a moderate worker, taking an excellent position in the classes, who, menstruation having appeared and revealed an error of sex, was suddenly removed by her parents; she afterwards became a beautiful woman and an excellent mother. There is even an instance told of a male pseudohermaphrodite brought up mistakenly as a woman, who became abbess of a convent and died in that dignified position at an advanced age. The Canoness Magdelena Mugnoz, after seven years in a cloister, was expelled because she was found to be of the male sex: and at the Convent des Filles de Dieu de Chartres, the Canoness Angélique de la Motte d'Aspermont, was accused of having played the man with her religious sisters, and the woman in her nocturnal escapades from the institution, and was deprived of her dignity as canoness and imprisoned" (Neugebauer, "Hermaphrodism in the Daily Practice," pp. 258–259). These stories also obviously contain morality plays about the proper roles of the sexes.

18. For cases of consultation regarding children, see, for example, Adrien de Mortillet, "Jeune hermaphrodite," *Bulletins de la société d'anthropologie de Paris,* 3d ser., 8 (1885): 650–652; H. Jolicoeur, "Hermaphrodisme," *Bulletin de la société médicale de Reims,* 12 (April 1873): 47–50; Passarini, "Pseudo-hermaphrodite androgynoïde," *Nouveau Montpellier médical,* 6 (1897): 354–358. For a case in which the parents tried to hide the children's ambiguity, see John Lindsay, "Three Cases of Doubtful Sex in One Family," *Glasgow Medical Journal,* 19 (1893): 161–165 (discussed in Chapter 3). Lawson Tait cataloged a number of "awkward mistakes" that arose when doubtful children were "referred to one sex or to the other by their parents without the direction of a skilled opinion"; Lawson Tait, *Diseases of Women* (New York: William Wood & Company, 1879), p. 21. For an account of the growth of medical authority in American obstetrics, see Judith Walzer Leavitt, "The Growth of Medical Authority: Technology and Morals in Turn-of-the-Century Obstetrics," *Medical Anthropology Quarterly,* 1 (1987): 230–255.

19. See, for instance, the case of Sarah Jane as reported in Jordan Lloyd, "Case of Spurious Hermaphroditism," *Illustrated Medical News,* 2 (1889): 103–104. Bynum notes that "the class and sex of the patient would have influenced how he or she was treated"; Bynum, *Science and the Practice of Medicine,* p. 177.

20. For "skilled opinion" issues see Tait, *Diseases of Women,* p. 21. The term "authorized interpreters" is from Xavier Delore, "Des Étapes de l'hermaphrodisme," *L'Écho médical de Lyon,* 4 (1899): 193–205, 225–232, p. 228. See also Granier, "Note sur un sujet atteint d'hypospadias pris jusqu'à 20 ans pour une femme; observations cliniques," *Nouveau Montpellier médical,* 3 (1894): 329–333, p. 331. There the author remarks on a particular case of mistaken sex: "It is very obvious that we have here a

case of very accentuated hypospadias, which at birth fooled the matrons who preside at births in the mountains of Brianconnais. An informed medical man would not have hesitated to recognize this deformity and assign the masculine sex."

21. Neugebauer was generally known as Franz in Germany and England and as François in France.

22. Krzysztof Boczkowski, "Franciszek Neugebauer (1856–1914): Pioneer in the Study of Hermaphroditism," *Polish Medical Science and History,* 9 (1966): 155–157.

23. Ibid., p. 156.

24. Ibid.

25. See, for example, Neugebauer, "Cinquante cas de mariages"; Franz Neugebauer, "Une Nouvelle série de 29 observations d'erreur de sexe," *Revue de gynécologie et de chirurgie abdominale,* 4 (1900): 133–174; Franz Neugebauer, "Quarante-quatre erreurs de sexe révélées par l'opération. Soixante-douze opérations chirurgicales d'urgence, de complaisance ou de complicité pratiquées chez des pseudo-hermaphrodites et personnes de sexe douteux," *Revue de gynécologie et de chirurgie abdominale,* 4 (1900): 457–518.

26. Lewis S. McMurty, "Jean Samuel Pozzi, M.D.," *Transactions of the American Gynecological Society,* 44 (1919): 383–384.

27. "Professor Pozzi," *Lancet,* 1 (June 22, 1918): 887–888.

28. McMurty, "Jean Samuel Pozzi," p. 383.

29. "Samuel Jean Pozzi, M.D.," *Boston Medical and Surgical Journal,* 179 (1918): 280–281.

30. See Samuel Pozzi, *A Treatise on Gynecology, Clinical and Operative* (London: New Sydenham Society, 1893).

31. "Décès de M. S. Pozzi," *Bulletin de l'académie de médecine,* 3d ser., 79 (1918): 448–452.

32. For a review of the complexities of medical specialization in nineteenth-century Paris, see George Weisz, "The Development of Medical Specialization in Nineteenth-Century Paris," in Ann LaBerge and Mordechai Feingold, eds., *French Medical Culture in the Nineteenth Century* (Atlanta: Rodopi, 1994).

33. See Samuel Pozzi, "Neuf cas personnels de pseudo-hermaphrodisme," *Revue de gynécologie et de chirurgie abdominale* 16 (1911): 269–339.

34. See, for example, Pozzi, "Neuf cas personnels"; Samuel Pozzi, "Homme hypospade considéré depuis vingt-huit ans comme femme (pseudo-hermaphrodite)," *Annales de gynécologie,* 21 (1884): 257–268; Samuel Pozzi, "Présentation d'un pseudo-hermaphrodite mâle," *Comptes rendus des séances de la société de biologie,* 8th ser., 1, pt. 2 (1884): 42–45; Samuel Pozzi, "De La Bride masculine du vestibule chez la femme et de l'origine de l'hymen," *Annales de gynécologie,* 21 (1884): 268–283; Samuel Pozzi, "Homme hypospade (pseudo-hermaphrodite) considéré depuis vingt-huit ans comme femme," *Bulletin de la société de médecine légale de France,* 8 (1885): 307–317; Samuel Pozzi, "Note sur deux nouveaux cas de pseudo-hermaphrodisme," *Gazette médicale de Paris,* 7th ser., 2 (1885): 109–112; Samuel Pozzi, "Note sur deux

nouveaux cas de pseudo-hermaphrodisme," *Comptes rendus des séances de la société de biologie,* 8th ser., 2, pt. 2 (1885): 23–29; Samuel Pozzi, "Pseudo-hermaphrodite mâle," *Bulletins de la société d'anthropologie de Paris,* 3d ser., 12 (1889): 602–608; Samuel Pozzi, "De L'Hermaphrodisme," *Gazette hebdomadaire de médecine et de chirurgie,* 2nd ser., 30 (1890): 351–355; Samuel Pozzi, "Sur un pseudo-hermaphrodite androgynoïde; prétendue femme ayant de chaque côté un testicule, un épididyme (ou trompe?) kystique et une corne utérine rudimentaire à gauche formant hernie dans le canal inguinal. Cure radicale. Examen microscopique," *Bulletin de l'académie nationale de médecine,* 3d ser., 36 (1896): 132–145; Samuel Pozzi, "Pseudo-hermaphrodite androgynoïde," *Presse médicale,* no. 64 (August 6, 1896): cccxiii–cccxv; Samuel Pozzi, "Sur une observation de M. le Dr. H. Barnsby (de Tours) intitulée: 'Pseudo-hermaphrodisme par hypospadias périnéo-scrotal,'" *Bulletins et mémoires de la société de chirurgie de Paris,* n.s., 32 (1906): 1103–1108.

35. "Professor Pozzi," p. 887.

36. "Jean Samuel Pozzi," *New York Medical Journal,* 108 (July 20, 1918): 121.

37. Dawson Williams, ed., "Robert Sydenham Fancourt Barnes," *British Medical Journal,* 1 (February 29, 1908): 541–542.

38. Thomas Wakley, ed., "Robert Sydenham Fancourt Barnes," *Lancet,* 1 (February 29, 1908): 683.

39. The rift was caused mainly by the contention among those who formed the Gynaecological Society that gynecology should form a special field of study, complete with its own hospitals and hospital wards, and especially that obstetricians and gynecologists, rather than general surgeons, should dominate female abdominal surgery. The Gynaecological Society members were particularly defensive of the right of obstetricians and gynecologists to perform lucrative ovariotomies; see Ornella Moscucci, *The Science of Woman: Gynaecology and Gender in England, 1800–1929* (New York: Cambridge University Press, 1990), pp. 171–173.

40. See, for example, Fancourt Barnes, "Spurious Hermaphroditism," *Transactions of the Obstetrical Society of London,* 24 (1882): 188–189; Fancourt Barnes, ed., "Ovariotomy on a Hermaphrodite [report of Dr. Florian Krug to the Gynaecological Society of Chicago, regular meeting of 20 Feb., 1891]," *British Gynaecological Journal,* 7 (1891–92): 254–255.

41. Robert Barnes and Fancourt Barnes, *A System of Obstetric Medicine and Surgery, Theoretical and Clinical, for the Student and the Practitioner,* 2 vols. (London: Smith, Elder & Co., 1884–85).

42. Dawson Williams, ed., "Lawson Tait," *British Medical Journal,* 1 (June 24, 1899): 1561–1564, p. 1562.

43. Leslie Stephen and Sidney Lee, "Lawson Tait," *Dictionary of National Biography* (1921–22), vol. 22, supplement, pp. 1249–50.

44. Williams, "Tait," p. 1561.

45. Ibid.

46. Stephen and Lee, "Tait," p. 1249.

47. Thomas H. Wakley and Thomas Wakley, eds., "Lawson Tait," *Lancet,* 1 (June 24, 1899): 1736–39, p. 1736.

48. Williams, "Tait," p. 1562.

49. See Moscucci, *Science of Woman,* p. 22.

50. Tait, *Diseases of Women,* p. 123.

51. See, for example, Chalmers, "Hermaphrodite [exhibited by Dr. Chalmers to the Obstetrical Society of London, meeting of 1 November, 1882]," *British Medical Journal,* 2 (November 18, 1882): 1003; and Walter E. Collinge, "On the Absence of Male Reproductive Organs in Two Hermaphrodite Molluscs," *Journal of Anatomy and Physiology,* 17 (1892–1893): 237–238, esp. p. 237.

52. For evolutionary arguments, see also Pozzi, "De L'Hermaphrodisme," p. 355; C. E. Underhill, "Case of Absence of Uterus, with a Tumour of Doubtful Character in Each Inguinal Canal," *Edinburgh Medical Journal* (1876): 906–913, p. 911; and Samuel Wilks, "Hermaphroditism," *British Medical Journal,* 2 (July 3, 1909): 48.

53. As quoted in Adrian Desmond and James Moore, *Darwin* (New York: Norton, 1991), p. 505.

54. See Lawson Tait, "Hermaphroditism," *Transactions of the American Gynecological Society,* 1 (1876): 318–325. Reprinted as "Hermaphroditism" (Boston, 1876).

55. See, for example, Tait, *Diseases of Women,* p. 22n.

56. See, for example, Gaffe, "Un Cas d'hermaphrodisme," *Journal de médecine et de chirurgie pratiques,* 56 (1885): 65–67, p. 67; Pozzi, "Homme hypospade" (1885), p. 317; and Pozzi, "De L'Hermaphrodisme," p. 351.

57. For instance, in the case of imperforation of the urethra, a surgeon might use what was known about hypospadias to guess where he would best try to create an opening; see L. Sentex, "Pseudo-hermaphrodisme apparent—Hypospadias péno-scrotal compliqué d'imperforation de l'urètre et d'absence des testicules," *Journal de médecine des Bordeaux et recueil des travaux de la société de médecine de la même ville,* 16 (1887): 54–55. Likewise, given a certain pattern of signs, a surgeon might suspect an apparent scrotum was really joined labia majora, in which case a well-placed slit might reveal an otherwise hidden vagina; see, for example, Pozzi, "De L'Hermaphrodisme," p. 352.

58. See, for example, John Alexander Smith, "Notice of True Hermaphrodism in the Codfish *(Morrhua vulgaris),* and in the Herring *(Clupea harengus),*" *Journal of Anatomy and Physiology,* 4 (1870): 256–258, which includes thanks to "Mr. W. Bargh, fishmonger, Earl Grey Street," who provided Smith with a hermaphroditic specimen of codfish, and to "Mr. Anderson, fishmonger, George Street," who provided a hermaphroditic herring. See also D. J. Cunningham, "A Hermaphroditic Goat," *Transactions of the Academy of Medicine, Ireland,* 3 (1885): 457–460, regarding the purchase of a hermaphroditic sheep from the farmer who brought it to his attention. See also Wilks, "Hermaphroditism," for a note on another hermaphroditic sheep provided to him for examination by his butcher.

59. See, for example, Neugebauer, "Cinquante cas de mariages," p. 202.

60. See Isidore Geoffroy Saint-Hilaire, *Histoire générale et particulière des anomalies de l'organization chez l'homme et les animaux . . . ou Traité de tératologie,* vol. 3 (Paris: J.-B. Baillière, 1832–1836), pp. 549ff.; see also Alice Domurat Dreger, "'Nature Is One Whole': Isidore Geoffroy Saint-Hilaire's *Traité de tératologie*" (M.A. thesis, Indiana University, 1993), pp. 149–160.

61. See Oscar Hertwig, *Text-Book of the Embryology of Man and Mammals,* trans. from the third edition by Edward L. Mark (London: Swan Sonnenschein & Company, 1892), p. 402.

62. See, for example, Hulke, "Four Cases of Congenital Defects of the Female Sexual Organs," *Lancet,* 1 (June 23, 1883): 1088–1089, p. 1088. For a discussion of Isidore Geoffroy Saint-Hilaire's influential early advocacy of this "law" of teratology, see Dreger, "'Nature Is One Whole,'" pp. 29–38.

63. Tait, *Diseases of Women,* p. 21; compare Pozzi, "De L'Hermaphrodisme," p. 351.

64. See, for example, Guermonprez, "Une erreur de sexe," p. 347, n. 1; and G. Lagneau, "Sur deux cas d'hermaphrodisme," *Bulletin de l'académie de médecine,* 3d ser., 33 (1895): 415–418.

65. Pozzi, "De La Bride masculin" (1884), p. 274; see also Dandois, "Un Exemple d'erreur de sexe par suite d'hermaphrodisme apparent," *Revue médicale,* 5 (1886): 49–52, p. 51; and Pozzi, "Homme hypospade" (1885), p. 307.

66. See, for example, Jouin, "Hermaphrodisme vrai et pseudo-hermaphrodisme," *Bulletins et mémoires de la société obstetricale et gynécologique de Paris* (1891): 190–197, p. 193.

67. Pozzi, "De La Bride masculin" (1884), p. 272. See also J. Jackson Clark, "A Case of Pseudo-Hermaphroditism," *Transactions of the Pathological Society of London,* 44 (1892–93): 120–122: "Here then is evidence that a considerable portion of the female urethra is formed from a region of the urogenital sinus, which is independent of the part utilised in the formation of the male urethra" (p. 122). See also comments of Duval in E. Magitot, "Sur un nouveau cas d'hermaphrodisme," *Bulletin de la société d'anthropologie de Paris,* 3d ser., 4 (1881): 487–496, p. 496; and p. 106 of Péan, "Hermaphrodisme masculin complexe. Arrêt de développement des organes génitaux mâles," *Gazette des Hopitaux civils et militaires,* 57 (1884): 105–107.

68. He did this essentially by evidencing the presence of the hymen in masculine pseudohermaphrodites who had vulvas and a small vagina but who lacked the superior part of the "true" vagina (see Pozzi, "De La Bride masculine").

69. Louis Guinard, "Un Cas d'hermaphrodisme parfait bisexuel," *Journal de médecine véterinaire et de zootechnie,* 3d ser., 15 (1890): 351–353, p. 352.

70. Jonathan Hutchinson, "A Note on Hermaphrodites," *Archives of Surgery,* 7 (1896): 64–66, p. 65.

71. J. W. Ballantyne, "Hermaphroditism," in Chalmers Watson, ed., *Encyclopaedia Medica,* vol. 4 (Edinburgh: William Green and Sons, 1900), p. 491. See also

G. Lombardi, "Contribution à l'étude de l'hermaphrodisme des voies génitales," *Presse médicale,* no. 53 (1905): 417–418; and Charles Debierre, "Note sur un merlan hermaphrodite," *Comptes rendus des séances de la société de biologie,* 8th ser., 4 (1887): 31–32, p. 31. For an extensive and very helpful review of concepts of latent hermaphroditism in Victorian England, see Ornella Moscucci, "Hermaphroditism and Sex Difference: The Construction of Gender in Victorian England," in Marina Benjamin, ed., *Science and Sensibility: Gender and Scientific Inquiry, 1780–1945* (Cambridge, Mass.: Basil Blackwell, 1991), pp. 174–199.

72. Neugebauer, "Hermaphrodism in the Daily Practice," p. 232.

73. Pozzi, "Homme hypospade" (1885), p. 316; see also Pozzi, "Sur un pseudo-hermaphrodite," p. 143. For a history of concepts of parental impressions, see Marie-Hélène Huet, *Monstrous Imagination* (Cambridge: Harvard University Press, 1993).

74. See, for example, the comments of Heywood Smith in Fancourt Barnes, [report of a living specimen of a hermaphrodite, meeting of 25 April 1888], *British Gynaecological Journal* 4 (1888): 205–212, p. 211; and see G. R. Green, "A Case of Hypospadias in a Patient, Aged 24, Who Had Always Passed as a Woman," *Quarterly Medical Journal* (Sheffield), 6 (1897–98): 130–132, p. 132.

75. Barnes, "Report of a Living Specimen," p. 205.

76. Pozzi, "Sur un pseudo-hermaphrodite," p. 134.

77. See, for example, Hart (report by Rigden), "Spurious Hermaphroditism," *British Medical Journal,* 2 (Dec. 2, 1888): 1214, where we read: "The mother . . . could not account for the deformity." See also Green, "A Case of Hypospadias," p. 132: "I communicated with the mother, but could elicit nothing from her. She tells me she had no shock or illness, and, in fact, had remarkably good health during that pregnancy." See also Jolicoeur, "Hermaphrodisme," p. 47.

78. Thomas H. Kellock, "Two Cases of Complete Hypospadias with Split Scrotum in Children of the Same Family," *Transactions of the Clinical Society of London,* 32 (1899): 242–243, p. 242.

79. Lindsay, "Three Cases," p. 163. Compare John Phillips, "Four Cases of Spurious Hermaphroditism in One Family," *Transactions of the Obstetrical Society of London,* 28 (1886): 158–168, esp. p. 159.

80. See, for example, Phillips, "Four Cases," p. 166; see also Alfred Lingard, "The Hereditary Transmission of Hypospadias and Its Transmission by Indirect Atavism," *Lancet,* 1 (April 19, 1884): 703.

81. See, for example, Arthur Maude, "A Case of Pseudo-Hermaphroditism," *British Gynaecological Journal,* 14 (1898): 429–432, p. 429.

82. See, for example, C. Stonham, "Case of Perfect Uterus Masculinus with Perfect Fallopian Tubes and Testes in the Broad Ligament. Complex or Vertical Hermaphroditism," *Transactions of the Pathological Society of London,* 29 (1887–88): 219–226.

83. See W. Bulloch, "Hereditary Malformation of the Genital Organs, Hermaphroditism," *Eugenics Laboratory Lecture Series,* Galton Laboratory for National Eugenics, University of London (1909): 50–61.

84. Henry Corby, "Removal of a Tumour from a Hermaphrodite," *British Medical Journal*, 2 (September 23, 1905): 710–712, p. 711.

85. For examples of investigations into hereditary antecedents, see Passarini, "Pseudo-hermaphrodite androgynoïde," pp. 357–358; Raoul Blondel, "Observation de pseudo-hermaphrodisme.—Un homme marié à un homme," *Journal de médecine de Paris*, 2d ser., 11 (1899): 75–77, p. 75; Martinet, "Un Cas d'hermaphrodisme," *Revue générale de clinique et de thérapeutique: Journal des praticiens*, 13 (1899): 402–404, p. 404; Ricoux and E. Aubry, "Un Prétendu androgyne dans un service de femmes," *Progrès médicale*, 3d ser., 10 (1899): 183–184, p. 183. For the question of consanguinity and "mental alienation," see, for example, Pozzi, "Note sur deux nouveaux cas," p. 110; Pozzi, "De L'Hermaphrodisme," p. 354; and Paul Petit, "Pseudo-hermaphrodisme périnéo-scrotal," *Bulletins et mémoires de la société obstetricale et gynécologique de Paris*, (1891): 130–131, p. 130. For issues of melancholia and alcoholism, see, for example, Guermonprez, "Une Erreur de sexe," p. 372, n. 1; and Arène, "Un Cas de pseudo-hermaphrodisme," p. 195. In his study of Louise Bavet, a male pseudohermaphrodite, Pozzi seemed surprised that "no hereditary antecedents could be incriminated" ("Homme hypospade" [1885], p. 315). The best explanation he could provide appealed to family lore, which had it that Louise was conceived after her parents had attended a wedding, at which "it was possible that a little more than reasonable might have been drunk" by her father (p. 316). Perhaps, then, the femininity of the truly-male Louise Bavet was due to a father's drunkenness on one critical night.

86. De Beurmann and Roubinovitch, "Pseudo-hermaphrodisme masculin (androgyne de Saint-Denis); présentation du sujet," *Bulletins et mémoires de la société médicale des hôpitaux de Paris*, 3d ser. 23 (1906): 47–58, p. 58.

87. See Sentex, "Pseudo-hermaphrodisme apparent," p. 55.

88. Jean Tapie, "Un Cas d'erreur sur le sexe. Malformation des organes génitaux externes; pathogénie de ces vices de conformation," *Revue médicale de Toulouse*, 22 (1888): 301–313, pp. 311–312.

89. For example, in 1884 Gérin-Roze suggested, as Geoffroy had, that if certain genital organs were robbed of nutritive blood supplies, they might suffer an arrest of development, and "make an effective male an apparent female"; Gérin-Roze, "Un Cas d'hermaphrodisme faux," *Bulletins et mémoires de la société médicale des hôpitaux*, 3d ser., 1 (1884): 369–373, p. 373.

90. See Neugebauer, "Hermaphrodism in the Daily Practice," p. 231. For more on Geoffroy, see Chapter 5.

91. In the late 1880s the French physiologist and neurologist Charles Edouard Brown-Séquard began publishing notices of his testicular extract experiments in the British and French medical literature; see Merriley Elaine Borell, "Organotherapy, British Physiology, and Discovery of the Internal Secretions," *Journal of the History of Biology*, 9 (1976): 235–268, p. 237. Additionally, many medical men had noted the "masculinization" that occurred in women whose ovaries were removed, and the "feminism" prevalent in men whose testicles remained undescended or atrophied.

92. Frank Lillie of the University of Chicago first made known his theory of the free-martin in the American journal *Science* in April of 1916; see Frank R. Lillie, "The Theory of the Free-Martin," *Science,* 43 (1916): 611–613; reprinted in Benjamin H. Willier and Jane M. Oppenheimer, eds., *Foundations of Experimental Embryology* (New York: Hafner Press, 1974), 138–142. Occasionally in cattle, twins are born in which one is a normal, functional male, and the other a hermaphroditic female, often infertile. "Free-martin" is, in English, the name given to the female twin. (The origin of the word is obscure.) Lillie employed a very clever combination of mathematics and embryology to provide evidence for his conviction that the free-martin was a true female made hermaphroditic through the influence of masculine hormones in embryonic life. Lillie was convinced that the free-martin was a true female whose ovaries developed to look extraordinarily testicular. (For further discussion, see Alice Domurat Dreger, "Doubtful Sex: Cases and Concepts of Hermaphroditism in France and Britain, 1868–1915" [Ph.D. dissertation, Indiana University, 1995], chap. 3. For a study of endocrinology and conceptualizations of human sex starting in 1920, see Nelly Oudshoorn, "Endocrinolgists and the Conceptualization of Sex, 1920–1940," *Journal of the History of Biology,* 23 [1990]: 163–186, and Nelly Oudshoorn, *Beyond the Natural Body: An Archeology of Sex Hormones* [New York: Routledge, 1994], chap. 2.) Lillie's was not the only theory of the free-martin at this time. In England in 1909, David Berry Hart (author of the "sex-ensemble" theory described in Chapter 5) proposed to the Royal Society of Edinburgh his own theory of the free-martin; see David Berry Hart, "The Structure of the Reproductive Organs in the Free-Martin, with a Theory of the Significance of the Abnormality," *Proceedings of the Royal Society of Edinburgh,* 30 (1909): 230–241. Berry Hart conducted both an extensive review of the literature on free-martins and a new examination of the preserved gonads from John Hunter's free-martin of some hundred and twenty years earlier. Certain that the free-martin was really a bull by virtue of its testes—he thought Hunter's specimen and the literature clearly showed that the sex glands of the free-martin were testes—Berry Hart grew equally convinced that the two twins must have originated from a single zygote. Characteristically for Berry Hart, his full theory of the etiology of the free-martin involved a complicated notion of Mendelian unit characteristics and the movement of those characteristics from cell to cell (see Hart, "The Structure," p. 238). Hart posited that the free-martin was a true male monozygotic twin whose fellow twin had been allotted the potent masculine Mendelian elements at the expense of the free-martin twin. Lillie's explanation ultimately won the day.

93. See, for example, T. W. P. Lawrence, "True Hermaphroditism in the Human Subject," *Transactions of the Pathological Society of London,* 57 (1905): 21–44.

94. P. Bouin and P. Ancel, "Sur un cas d'hermaphrodisme glandulaire chez les mammifères," *Comptes rendus des séances de la société de biologie,* 57 (1904): 656–657, p. 657.

95. Samuel G. Shattock and C. G. Seligmann, "An Example of True Hermaphroditism in the Domestic Fowl, with Remarks on the Phenomenon of Allopterotism,"

Transactions of the Pathological Society of London, 57 (1906): 69–109. This is the same Seligmann who, in 1902, authored a review article entitled "Sexual Inversion [i.e., homosexuality] among the Primitive Races" (see *Alienist and Neurologist,* 23 [1902]: 11–14). Many of the investigators interested in sex and hermaphroditism were also interested in sexuality and especially homosexuality (see Chapter 4).

96. Shattock and Seligmann, "An Example of True Hermaphroditism," p. 71.

97. S. G. Shattock and C. G. Seligmann, "An Example of Incomplete Glandular Hermaphroditism in the Domestic Fowl," *Proceedings of the Royal Society of Medicine,* Pathology Section, 1 (1907–1908): 3–7, p. 7; original emphasis.

98. See, for example, Pozzi, "Neuf cas personnels." In this review article on cases of human hermaphroditism, Samuel Pozzi insisted that "subjects who carry this [ovotesticular] organ always have a very characteristic feminine appearance and the ovarian functions sometimes do not even appear to be troubled" (p. 270). See also Chapter 5.

99. See P. Delagénière, "Anomalies des organes génitaux," *Annales de gynécologie et d'obstetrique,* 51 (1899): 57–63, p. 62.

100. Wilks, "Hermaphroditism." See also Samuel Wilks, "Hermaphroditism," *British Medical Journal,* 1 (June 4, 1910): 1377.

101. On castration, see also A. Pittard, "La Castration chez l'homme et les modifications qu'elle apporte [presentée par A. Laveran]," *Annales des maladies des organes génito-urinaires,* 1 (1904): 233–235. Wilks also related a case of a sheep he himself had seen: "My butcher brought me one day a small body which he found on a leg of mutton which he had cut from a ewe. It turned out to be a testis, and the so-called ewe was an ill-developed ram"; from Wilks, "Hermaphroditism" (1909). Albert Churchward of London was inspired by Wilks's 1909 letter in the *British Medical Journal* to write and offer the corroborating evidence of a case in his own practice in which the testicles had not descended and the result had been a womanly man; see Albert Churchward, "Hermaphroditism," *British Medical Journal,* 2 (August 14, 1909): 421.

102. These were probably cases of what today would be labeled testicular feminization syndrome, which, as noted in Chapter 1, is now considered a problem of hormone receptors, not hormone production. Without a clear concept of hormone physiology or a notion of receptors, it was virtually impossible to fit these sorts of female pseudohermaphrodites into an endocrinological theory of sexual development.

103. So, Neugebauer repeatedly insisted, "pseudo-hermaphroditism in its essence . . . is a paradoxical incongruity between the genital *glands* on the one hand and the genital *ducts* and external genitals on the other"; Bulloch, "Hereditary Malformation," p. 50, paraphrasing Franz Neugebauer, *Hermaphroditismus beim Menschen* (Leipzig: Werner Klinkhardt, 1908); original emphasis in Bulloch. See also Pozzi, "Neuf cas personnels," p. 273.

104. See Pozzi, "Neuf cas personnels," p. 284.

105. J. Hammond, "A Case of Hermaphroditism in the Pig," *Journal of Anatomy and Physiology,* 46 (1912): 307–312, p. 312.

106. See, for example, William Blair Bell, "Hermaphroditism," *Liverpool Medico-Chirurgical Journal,* 35 (1915): 272–292, pp. 289–290, and Chapter 5. This possibility in turn suggested that perhaps injection or ingestion of extracts from healthy glands, or transplantation of those healthy glands into the hermaphrodite, might serve as a form of treatment for masculinization in women and feminization in men; see, for example, the recommendation of Apert on page 138 in L. Guinon and Bijon, "Déviation du type sexuel chez une jeune fille, caractérisé par l'obésité et le développement d'attributs masculins simulant l'hermaphrodisme," *Bulletins de la société de pédiatrie de Paris,* 8 (1906): 129–138. As noted in William Blair Bell's case of S.B. in Chapter 5, early twentieth-century medical men sometimes tried to use ovarian transplants to alleviate the onset of masculine characteristics in a woman; similarly, at least in Germany testicular transplants were shown to "cure" homosexuality in men in the 1910s (see Chandak Sengoopta, "Glandular Politics: Endocrinology, Sexual Orientation, and Emancipation in Early Twentieth Century Central Europe," forthcoming). Cases of human hermaphroditism did not shed any light on the role of the chromosomes in the development of human sex during the period in question here. Chromosomes of human hermaphroditic subjects could not be examined until Barr body tests became available in the 1950s, and, even if the tests had been developed sooner, they would not have aided in the development of the chromosomal theory of sex (the theory of the XX and XY chromosomal bases of sex), because the "sex chromosomes" of many hermaphrodites microscopically appear to be like the typical male's or typical female's. (See Chapter 1.) For a history of sex determination theories from 1880 to 1916 with a special focus on chromosomes, see Jane Maienschein, "What Determines Sex? A Study of Converging Approaches, 1880–1916," *Isis,* 75 (1984): 457–480.

107. For more on this, see Chapter 4.

108. Neugebauer, "Hermaphrodism in the Daily Practice," p. 238.

109. See, for example, ibid., p. 239: "Pseudo-hermaphrodites have in numerous instances been in difficulties with the various authorities—ecclesiastical and judicial tribunals, magistrates, police or gendarmerie, and even with the directors of schools. Before the ecclesiastical courts they have figured in numerous cases for divorce; before the civil law the most common accusations against them have been for rape, seduction followed by pregnancy, unnatural crimes, and even sodomy. There have been also cases of murder committed by pseudo-hermaphrodites, and in many instances they have been the victims of crimes committed by others." Compare Leblond, "Du Pseudo-hermaphrodisme," p. 294: "One understands how errors [of sex] are injurious to individuals carrying these congenital malformations, which oblige them to live in the company of individuals of a different sex and to wear the clothes which are not agreeable to their aptitudes and their tastes."

110. Tapie, "Un Cas d'erreur," p. 302.

111. Ibid.

112. Tapie had recently completed an extensive study of the deformities of the hand.

113. Tapie, "Un Cas d'erreur," p. 303.

114. Ibid., p. 305.

115. Ibid., p. 303.

116. Ibid., p. 304.

117. Ibid., p. 306.

118. Ibid.

119. Ibid.

3. In Search of the Veritable Vulva

1. Franz Neugebauer, "Hermaphrodism in the Daily Practice of Medicine; Being Information upon Hermaphrodism Indispensable to the Practitioner," *British Gynaecological Journal,* 29 (1903): 226–263, p. 226.

2. Ibid., p. 246.

3. J. Halliday Croom, "Two Cases of Mistaken Sex in Adult Life," *Transactions of the Edinburgh Obstetrical Society,* 24 (1898–99): 102–104, p. 102.

4. Ibid.

5. Ibid.

6. Ibid., p. 103.

7. Ibid., p. 104.

8. Ibid.

9. Ernest Hart, ed., "Sexual Ignorance," *British Medical Journal,* 2 (Aug. 15, 1885): 303–304, p. 303.

10. Lawson Tait, *Diseases of Women* (New York: William Wood & Company, 1879), p. 21. Neugebauer claimed that "Lawson Tait advised that it [the child of doubtful sex] should be treated as a girl on the ground that it is far easier to protect a child from the disagreeable results of a genital malformation in the social position of a girl than in that of a boy" (Neugebauer, "Hermaphrodism in the Daily Practice," p. 244). However, I have not found recorded any such recommendation by Tait.

11. Pierre Garnier, "Du Pseudo-hermaphrodisme comme impédiment médico-légal a la déclaration du sexe dans l'acte de naissance," *Annales d'hygiene publique et de médecine légale,* 3d ser., 14 (1885): 285–293, p. 288.

12. G. R. Green, "A Case of Hypospadias in a Patient, Aged 24, Who Had Always Passed as a Woman," *Quarterly Medical Journal* (Sheffield), 6 (1897–98): 130–132, p. 131. See Chapter 4 for a longer discussion of this case. For group consultation, see also George Buchanan, "Hermaphrodite, Aged 9, in Whom Two Testicles Were Excised from the Labia Majora," *Glasgow Medical Journal,* 23 (1885): 213–217: "So enigmatical was the case that Dr. Lesichman and Dr. Cleland, as well as members of the Infirmary staff, were consulted, without absolute certainty as to diagnosis" (p. 214).

13. See Tait, *Diseases of Women,* p. 22n.

14. See De Gorrequer Griffith, "A Case of Hermaphroditism," *Proceedings of the Medical Society of London,* 3 (1875): 25–27.

15. Brian Rigden, "Spurious Hermaphroditism [described by Mr. Brian Rigden to the South-Eastern Branch, East Kent District]," *British Medical Journal,* 2 (December 2, 1882): 1119.

16. Alcide Benoist, "Rapport sur un cas d'hermaphrodisme," *Annales d'hygiene publique et de médecine légale,* 3d ser., 16 (1886): 84–87, p. 87. Descoust brought another case before the same society that same year; see Descoust, "Sur un cas d'hermaphrodisme," *Annales d'hygiene publique et de médecine légale,* 3d ser., 16 (1886): 87–90.

17. J. W. Ballantyne, "Hermaphroditism," in Chalmers Watson, ed., *Encyclopaedia Medica,* vol. 4 (Edinburgh: William Green and Sons, 1900), p. 490.

18. Theophilus Parvin, *The Science and Art of Obstetrics* (Philadelphia: Lea Brothers and Company, 1886), p. 84, original emphasis.

19. Tait, *Diseases of Women,* p. 124.

20. Frederick Willcocks, "A Case of Hypospadias (Mistaken Sex)," *Transactions of the Pathological Society of London,* 36 (1884–85): 309–311, p. 310.

21. Samuel G. Shattock, "A Specimen of Incomplete 'Transverse' Hermaphroditism in the Female, with a Note on the Male Hymen," *Transactions of the Pathological Society of London,* 41 (1889–90): 196–201, p. 196.

22. Edis quoted in Fancourt Barnes, [report of a living specimen of a hermaphrodite, meeting of 25 April 1888], *British Gynaecological Journal* 4 (1888): 205–212, p. 212.

23. John Lindsay, "Three Cases of Doubtful Sex in One Family," *Glasgow Medical Journal,* 19 (1893): 161–165, p. 164. Neugebauer believed that "in the majority of cases the true sex, even when indeterminate at birth, declares itself spontaneously at puberty" (Neugebauer, "Hermaphrodism in the Daily Practice," p. 242).

24. Arène, "Un Cas de pseudo-hermaphrodisme," *Loire médicale,* 14 (1895): 187–195, p. 193.

25. Hemey and Gallard quoted in Descoust, "Sur un cas d'hermaphrodisme," p. 90.

26. C. E. Underhill, "Case of Absence of Uterus, with a Tumour of Doubtful Character in Each Inguinal Canal," *Edinburgh Medical Journal* (April 1876): 906–913, p. 908.

27. See, for example, De Gorrequer Griffith, "A Case of Hermaphroditism," p. 26.

28. Neugebauer, "Hermaphrodism in the Daily Practice," p. 250; see also Andrew Clark, "A Case of Spurious Hermaphroditism (Hypospadias and Undescended Testes in a Subject Who Had Been Brought Up as a Female and Been Married for Sixteen Years)," *Lancet* 1 (March 12, 1898): 718–719, esp. p. 719.

29. Neugebauer, "Hermaphrodism in the Daily Practice," p. 230; see also Samuel Pozzi, "De L'Hermaphrodisme," *Gazette hebdomadaire de médecine et de chirurgie,* 2d ser., 27 (1890): 351–355, esp. pp. 351–352.

30. Lindsay, "Three Cases of Doubtful Sex," p. 163.

31. Ibid.

32. Barnes, "Report of a Living Specimen," p. 206.

33. Lindsay, "Three Cases of Doubtful Sex," p. 164.

34. Edis quoted in Barnes, "Report of a Living Specimen," p. 212.

35. Lindsay, "Three Cases of Doubtful Sex," p. 164.

36. Arthur Maude of Westerham used this statistical reasoning when he argued that although "[t]he rule in doubtful cases used to be to have the child brought up as a female, on the ground that concealment of abnormality was easier for a girl than a boy," in fact "this rule is a bad one in face of the unexplained fact of the far greater excess of masculine pseudo-hermaphrodites over female." Maude also suggested it was better to raise doubtful children as males because of "the greater social difficulties which, at a marriageable age, will await the male in petticoats than the female in trousers." (See Arthur Maude, "A Case of Pseudo-Hermaphroditism," *British Gynaecological Journal,* 14 [1898]: 429–432, p. 431.) On the issue of the relative frequency of pseudohermaphroditism in males versus females, see also Barton Cooke Hirst and George A. Piersol, *Human Monstrosities* (Edinburgh and London: Young J. Pentland, 1892; in four parts), see esp. p. 80. See also Clark, "A Case of Spurious Hermaphroditism," p. 718; and G. R. Turner, "A Case of Hermaphroditism," *Lancet,* 1 (June 30, 1900): 1884–85, esp. p. 1884. For another warning about the preponderance of male pseudohermaphrodites, see Xavier Delore, "Des Étapes de l'hermaphrodisme," *L'Écho médical de Lyon,* 4, no. 7 (1899): 193–205 and no. 8 (1899): 225–232: "But, moreover, one should not forget that most hermaphrodites are of the masculine genre; that the etiogenical mode . . . explains sufficiently their numerical predominance; that all the judiciary dramas have had at their origin mistaken males" (p. 228).

37. See, for example, Green, "A Case of Hypospadias," p. 130: "in voice and outward appearance she is apparently a woman." See also Samuel Pozzi, "Pseudo-hermaphrodite mâle," *Bulletins de la société d'anthropologie de Paris,* 3d ser., 12 (1889): 602–608: "The voice is nearly feminine; the larynx is undeveloped; the Adam's apple scarcely accentuated" (p. 602).

38. Barnes, "Report of a Living Specimen," p. 212.

39. Granier, "Note sur un sujet atteint d'hypospadias pris jusqu'à 20 ans pour une femme; observations cliniques," *Nouveau Montpellier médical,* 3 (1894): 329–333, p. 330.

40. See Croom, "Two Cases," p. 102; see also Robert Barnes and Fancourt Barnes, *A System of Obstetric Medicine and Surgery, Theoretical and Clinical, for the Student and Practitioner,* 2 vols. (London: Smith, Elder & Co., 1884–85), vol. 1, p. 531.

41. See Gaffe, "Un Cas d'hermaphrodisme," *Journal de médecine et de chirurgie pratiques,* 56 (February 1885): 65–67, p. 65. Compare A. Lutaud, "De L'Hermaphrodisme au point de vue médico-légal. Nouvelle observation. Henriette Williams," *Journal de médecine de Paris* (1885): 386–396: "[T]he voice sweet and agreeable and of a rather feminine timbre" (p. 392).

42. See, for example, Accolas, "Cas de pseudo-hermaphrodisme," *Revue médico-chirurgical des maladies de femmes,* 11 (1889): 140–144: "The voice is rather grave; the

timbre differs a little from that of a contralto, and approaches more that of a man; but it is sweet, almost tender" (p. 141).

43. See, for example, Samuel Pozzi, "Note sur deux nouveaux cas de pseudo-hermaphrodisme," *Gazette médicale de Paris,* 7th ser., 2 (1885): 109–112, esp. p. 109.

44. Jean Tapie, "Un Cas d'erreur sur le sexe. Malformation des organes génitaux externes; pathogénie de ces vices de conformation," *Revue médicale de Toulouse,* 22 (1888): 301–313, p. 303.

45. Barnes, "Report of a Living Specimen," p. 206.

46. Brouardel, "Hermaphrodisme; impuissance; type infantile," *Gazette des hôpitaux civils et militaires,* 60 (1887): 57–59, p. 57.

47. Delore, "Des Étapes," p. 228.

48. Albert Leblond, "Du Pseudo-hermaphrodisme comme impédiment médico-légal a la déclaration du sexe dans l'acte de naissance," *Annales de gynécologie,* 24 (1885): 25–36, p. 26.

49. Garnier, "Du Pseudo-hermaphrodisme," p. 291.

50. Arnold Davidson relies on the *Oxford English Dictionary* to trace back to 1879 the appearance of the modern usage of "sexuality" ("possession of sexual powers, or capability of sexual feeling") in the English language. (See Arnold Davidson, "Sex and the Emergence of Sexuality," in Edward Stein, ed., *Forms of Desire: Sexual Orientation and the Social Constructionist Controversy* [New York: Routledge, 1990], pp. 89–132.) Davidson marks the shift in an illustrative passage provided by the *O.E.D.* and drawn from J. M. Duncan's *Diseases of Women* of 1879: "In removing the ovaries, you do not necessarily destroy sexuality in a woman" (quoted on pp. 98–99 in Davidson). While I do not disagree with Davidson's general point (that the modern idea of sexuality as distinct from sex began to emerge only in the late nineteenth century), unfortunately the *O.E.D.* and Davidson together have assumed that Duncan meant what the late twentieth century means by "sexuality" and have therefore misinterpreted Duncan's phraseology. Duncan almost certainly did *not* draw a sex/sexuality distinction, nor did his colleagues; instead they used the broad term "sexuality" to denote all those characteristics (anatomical features, behaviors, penchants, talents, and abilities) typically associated with one or the other sex. Even as late as 1899 the *New Sydenham Society's Lexicon of Medicine and the Allied Sciences* defined "sexuality" as broadly and vaguely as Duncan probably would have: "the characteristics of sex; those special characters which go to constitute either a male or a female." When Duncan said a woman whose ovaries were removed did not necessarily lose her sexuality, he thus most likely meant that she did not necessarily lose her "femininity," that is, her basically "feminine" anatomy, tastes, talents, and so on. Ironically, in his remark, Duncan was actively demonstrating the lack of a distinction between sex and sexuality that Davidson assumes he displayed. For more on what we call sexuality, see Chapter 4.

51. Barnes, "Report of a Living Specimen," p. 205.

52. See Péan, "Sur un cas d'hermaphrodisme," *Bulletin de l'académie de médecine,* 3d ser., 33 (1895): 381–383: "Some years later, he [a girl made boy] showed so little

aptitude for the masculine work [given to him] . . . that Dr. Mozer convened anew some doubts about his sex."

53. Maude, "A Case of Pseudo-Hermaphroditism," p. 429.

54. Pozzi, "Pseudo-hermaphrodite," p. 606. Compare Tapie, "Un Cas d'erreur," on an adolescent who "excels, according to her mother, in certain works of embroidery which are not lacking in difficulty. When she abandons her ordinary occupations, she delights in . . . [such activities] as striking a hammer, playing with whips, and so on."

55. See, for example, Tapie's claims as recounted in Chapter 2.

56. See, for example, in Gérin-Roze, "Un Cas d'hermaphrodisme faux," *Bulletins et mémoires de la société médicale des hôpitaux,* 3d ser., 1 (1884): 369–373; note p. 370: "Julie has . . . the timidity of a young woman." Note also that A.B., the menstruating man, had a "shy, cowed look" so unmanly in William Rushton Parker, "A Menstruating Man: A Curious Form of Hermaphroditism," *British Medical Journal,* 1 (February 4, 1899): 272.

57. Jonathan Hutchinson, "Gynaecomazia and Other Aberrations in the Development of Sex," *Archives of Surgery,* 3 (April 1892): 327–331, p. 328.

58. Granier, "Note sur un sujet," p. 331. Compare the description of Pauline S. as "anything but shy and reserved; the character was decidedly masculine" in J. H. Targett, "Two Cases of Pseudo-Hermaphroditism," *Transactions of the Obstetrical Society of London,* 36 (1895): 272–276, p. 274.

59. Neugebauer, "Hermaphrodism in the Daily Practice," p. 235.

60. See, for example, Barnes, "Report of a Living Specimen," p. 205, and see also Targett, "Two Cases."

61. Lutaud, "De L'Hermaphrodisme," p. 394; compare the description of the "bearded woman," Emilie M., in Jablonsky, "Note sur un cas d'hermaphrodisme," *Le Poitou médical,* 11 (1897): 274–276, on p. 274.

62. Brouardel, "Hermaphrodisme," p. 57. "Moral orientation" could here also be translated as "intellectual orientation"; the French word *(morale)* carries both meanings.

63. John Phillips, "Four Cases of Spurious Hermaphroditism in One Family," *Transactions of the Obstetrical Society of London,* 28 (1886): 158–168, p. 159.

64. C. Stonham, "Case of Perfect Uterus Masculinus with Perfect Fallopian Tubes and Testes in the Broad Ligament. Complex or Vertical Hermaphroditism," *Transactions of the Pathological Society of London,* 29 (1887–88): 219–226, p. 223. On the French practice of seeking family histories in the search for causes of "degeneracy," see Chapters 2 and 4; on the taking of sexual histories, see Chapter 4.

65. See, for example, Croom, "Two Cases," p. 102; the remarks of Aveling in Barnes, "Report of a Living Specimen"; and Polaillon, "Un Cas d'hermaphrodisme," *Bulletins et mémoires de la société obstetricale et gynécologique de Paris,* 2 (1887): 162–165, esp. p. 163. For a discussion of the tricky nature of gynecological exams during the Victorian period, see Mary Poovey, "'Scenes of an Indelicate Character': The Medical 'Treatment' of Victorian Women," *Representations,* 14 (1986): 137–168.

66. See, for instance, Accolas, "Cas de pseudo-hermaphrodisme"; Polaillon, "Un Cas d'hermaphrodisme," p. 163; Green, "A Case of Hypospadias," p. 131; Pozzi, "Note sur deux nouveaux cas," p. 111; and Christopher Martin, "A Case of Hermaphroditism," *British Gynaecological Journal,* 10 (1895): 35–37.

67. See Barnes, "Report of a Living Specimen," p. 206; Willcocks, "A Case of Hypospadias"; Martin, "A Case of Hermaphroditism"; Rigden, "Spurious Hermaphroditism"; and Jordan Lloyd, "Case of Spurious Hermaphroditism," *Illustrated Medical News,* 2 (1889): 103–104.

68. See, for example, Lloyd, "Case of Spurious Hermaphroditism"; Martin, "A Case of Hermaphroditism"; and Rigden, "Spurious Hermaphroditism."

69. Barnes, "Report of a Living Specimen," p. 212.

70. See, for example, Lindsay, "Three Cases," p. 164.

71. Barnes, "Report of a Living Specimen," p. 211.

72. Ibid., p. 210; see also Lloyd, "Case of Spurious Hermaphroditism"; Green, "A Case of Hypospadias," p. 131; Pozzi, "Pseudo-hermaphrodite," p. 603.

73. Stéphane Bonnet and Paul Petit, *Traité pratique de gynécologie* (Paris: J.-B. Baillière et Fils, 1894), pp. 74–75.

74. Delore, "Des Étapes," p. 230.

75. See François Jules Octave Guermonprez, "Une Erreur de sexe, avec ses conséquences," *Journal des sciences médicales de Lille* 2 (1892): 337–349, 361–376, p. 343.

76. Even biopsies, once they were used in the second decade of the twentieth century, seemed sometimes only to add to the confusion; see, for example, the case of S.B. summarized in Chapter 5.

77. Péan, "Hermaphrodisme," p. 383.

78. Buchanan, "Hermaphrodite, Aged 9," p. 214.

79. See, for example, Martin, "A Case of Hermaphroditism"; see also P. Delagénière, "Anomalies des organes génitaux," *Annales de gynécologie et d'obstetrique,* 51 (1899): 57–63.

80. Samuel Pozzi, "Sur un pseudo-hermaphrodite androgynoïde; prétendue femme ayant de chaque côté un testicule, un épididyme (ou trompe?) kystique et une corne utérine rudimentaire à gauche formant hernie dans le canal inguinal. Cure radicale. Examen microscopique," *Bulletin de l'académie nationale de médecine,* 3d ser., 36 (1896): 132–145, pp. 133–134. Compare Stonham, "Case of Perfect Uterus Masculinus," in which surgery was performed to repair a hernia in a child thought unequivocally male: "On August 5th, 1885, the child was put under chloroform, and the operation conducted in the usual way. On opening the sac it was found to contain what seemed to be a uterus with a broad ligament stretching off from each side, and also two organs properly placed for ovaries, and what seemed to be Fallopian tubes" (p. 219).

81. Clark, "A Case of Spurious Hermaphroditism," pp. 718–719; see also Martin, "A Case of Hermaphroditism," p. 36; and Phillips, "Four Cases," p. 160.

82. See, for example, Samuel Pozzi, "Présentation d'un pseudo-hermaphrodite male," *Comptes rendus des séances de la société de biologie,* 8th ser., 1, pt. 2 (1884): 42–45, p. 43; A. Primrose, "A Case of Uterus Masculinus—Tubular Hermaphroditism—in the Male, with Sarcomatous Enlargement of an Undescended Testicle," *Journal of Anatomy and Physiology,* 33 (1898–99): 64–75; and Pozzi, "Pseudo-hermaphrodite." Medical practitioners also asked doubtful patients with scrotal and labial lumps to cough while the examiner felt the swelling; this gave clues as to whether or not the lumps were indeed testicles or herniated organs, since a cough would increase the pressure on a hernia. Signs of particular cords (such as seminal vesicles) that would be associated with a particular sort of gonad were also sought.

83. Descoust, "Sur un cas d'hermaphrodisme," p. 88.

84. Green, "A Case of Hypospadias," p. 131.

85. See, for example, Thomas H. Kellock, "Two Cases of Complete Hypospadias with Split Scrotum in Children of the Same Family," *Transactions of the Clinical Society of London,* 32 (1899): 242–243; Guermonprez, "Une Erreur de sexe," p. 342; and Gérin-Roze, "Un Cas d'hermaphrodisme faux," p. 370.

86. Buchanan, "Hermaphrodite, Aged 9," p. 217.

87. See Neugebauer, "Hermaphrodism in the Daily Practice," p. 246.

88. See, for example, H. Cooper Rose, "Case of Extreme Hypospadias in a Child Who Had Been Baptised, Brought Up, and Educated at a Large Girls' School as a Female," *Transactions of the Obstetrical Society of London,* 18 (1877): 256–258, which reports a patient in whom "the slightest irritation [produced] a firm erection" (p. 257).

89. See Accolas, "Cas de pseudo-hermaphrodisme"; Gérin-Roze, "Un Cas d'hermaphrodisme faux," p. 372; see also Arène, "Un Cas de pseudo-hermaphrodisme."

90. Lutaud, "De L'Hermaphrodisme," p. 395. On the history of women's orgasms, see Thomas Laqueur, "Orgasm, Generation, and the Politics of Reproductive Biology," in *Representations,* 14 (1986): 1–41.

91. See, for example, p. 416 in G. Lagneau, "Sur deux cas d'hermaphrodisme," *Bulletin de l'académie de médecine,* 3d ser., 33 (1895): 415–418; see also Lindsay, "Three Cases," p. 162.

92. See the remarks of Heywood Smith in Barnes, "Report of a Living Specimen," p. 211; see also Adrien Pozzi and P. Grattery, "Pseudo-hermaphrodisme (hypospadias périnéal)," *Le Progrès médical,* 2d ser., 5 (1887): 308–312, p. 308.

93. See the remarks of Edis in Barnes, "Report of a Living Specimen," p. 212.

94. See, for instance, Pozzi and Grattery, "Pseudo-hermaphrodite," p. 309; Tapie, "Un Cas d'erreur," pp. 303–304; Targett, "Two Cases," p. 275; and H. Jolicoeur, "Hermaphrodisme," *Bulletin de la société médicale de Reims,* 12 (April 1873): 47–50.

95. Barnes, "Report of a Living Specimen," p. 212. See also J. Jackson Clarke, "A Case of Pseudo-Hermaphroditism," *Transactions of the Pathological Society of London,* 44 (1892–93): 120–122, for a report of "a child christened Frances on account of the sex being unknown" having "a perforate penis and urethra with the male characters" (p. 120).

96. See, for example, John Wood, "The Pelvis and Genital Organs of an Adult Hermaphrodite," *Transactions of the Pathological Society of London,* 23 (1872): 169–175, p. 170

97. See, for example, Chalmers, "Hermaphrodite [exhibited by Dr. Chalmers to the Obstetrical Society of London, meeting of 1 November, 1882]," *British Medical Journal,* 2 (November 18, 1882): 1003; and Fancourt Barnes, "Spurious Hermaphroditism," *Transactions of the Obstetrical Society of London,* 24 (1882): 188–189, p. 188.

98. For "a clitoris, or more justly a penis having a length of about 0.03 cm. [*sic*], supplied with a prepuce, a meatus, and a canal giving passage to the urine, all well formed, despite the smallness," see Thezet, "Un Cas singulier d'anomalie des organes génito-urinaires," *Union médicale,* 22 (1876): 612–613, p. 612.

99. On the expectation that clitorises would be smaller than penises, see, for example, Phillips, "Four Cases," p. 162, and Shattock, "A Specimen of Incomplete 'Transverse' Hermaphroditism"; on enlarged clitorises and atrophied penises, see Tapie, "Un Cas d'erreur," p. 304.

100. See, for example, p. 415 in Lagneau, "Sur deux cas."

101. Neugebauer, "Hermaphrodism in the Daily Practice," p. 255.

102. For a typically vague British report, see Wood, "The Pelvis and Genital Organs"; see also Sir Hector Clare Cameron, "Notes on a Case of Hermaphrodism," *British Gynaecological Journal,* 29 (1904): 347–351, p. 350.

103. See, for instance, Clark, "A Case of Spurious Hermaphroditism," p. 719; and Richard Boddaert, "Étude sur l'hermaphrodisme latéral," *Annales de la société de médecine de Gand,* 52 (1874): 81–139.

104. Raoul Blondel, "Observation de pseudo-hermaphrodisme.—Un homme marié à un homme," *Journal de médecine de Paris,* 2d ser., 11 (1899): 75–77, p. 76.

105. Guermonprez, "Un Cas d'erreur," p. 367.

106. See, for example, E. Magitot, "Sur un nouveau cas d'hermaphrodisme," *Bulletin de la société d'anthropologie de Paris,* 3d ser., 4 (1881): 487–496, p. 491. For similar "nocturnal pollutions," see Pozzi, "Présentation d'un pseudo-hermaphrodite," p. 44.

107. For a case in which the semen was believed to be devoid of spermatozoa, see Pozzi, "Pseudo-hermaphrodite," p. 607.

108. Arène, "Un Cas de pseudo-hermaphrodisme," p. 193.

109. On the absence as a present sign see, for example, Guermonprez, "Un Cas d'erreur" (p. 344), quoting Garnier, "Du Pseudo-hermaphrodisme."

110. Neugebauer, "Hermaphrodism in the Daily Practice," p. 252.

111. Ibid., p. 252. Contrary to this claim by Neugebauer, my work indicates it was more likely supposed hernias in supposed women which turned out to be recently descended testicles that "more frequently than any other [condition] . . . led to the discovery of an error of sex."

112. See comments of Edis in Barnes, "Report of a Living Specimen," p. 212, and see Primrose, "A Case of Uterus Masculinus." See also Targett, "Two Cases," p. 276.

113. See, for instance, Granier, "Note sur un sujet," p. 329, and Martin, "A Case of Hermaphroditism," p. 35.

114. For a case in which medical men found ovaries but no menses, see Bacaloglu and Fossard, "Deux cas de pseudo-hermaphroditisme (gynandroïdes)," *Presse Médicale,* 2 (1899): 331–333, p. 333.

115. Pozzi, "Pseudo-hermaphrodite," p. 608.

116. Ibid.

117. See, for example, Stonham, "Case of Perfect Uterus Masculinus," p. 226.

118. Guermonprez, "Un Cas d'erreur," p. 367, n. 1, with original emphasis. See also Benoist, "Rapport sur un cas d'hermaphrodisme," p. 85; Péan, "Sur un cas d'hermaphrodisme," p. 383; Pozzi, "Sur un pseudo-hermaphrodite," p. 134; Barnes, "Report of a Living Specimen," pp. 206, 212; Underhill, "Case of Absence of Uterus," pp. 907, 912. On the issue of Victorian menstruation, see also Sally Shuttleworth, "Female Circulation: Medical Discourse and Popular Advertising in the Mid-Victorian Era," in Mary Jacobus, Evelyn Fox Keller, and Sally Shuttleworth, eds., *Body/Politics: Women and the Discourses of Science* (New York: Routledge: 1990), pp. 47–68.

119. Cameron, "Notes on a Case of Hermaphrodism," p. 349.

120. Clark, "A Case of Spurious Hermaphroditism," p. 718; see also Neugebauer, "Hermaphrodism in the Daily Practice," p. 256.

121. Ahfeld as cited in Samuel Pozzi, *A Treatise on Gynecology, Clinical and Operative,* 3 vols. (London: New Syndenham Society, 1893), p. 464n. It was therefore best to have claims verified by a doctor, as in Magitot, "Sur un nouveau cas": "M. Foley asked if the appearance of the menses in this individual had been established by a doctor" (p. 492). Magitot replied, "No. We are reduced on this point to the information which has been furnished to us by the subject" (p. 493).

122. For example, in Accolas, "Cas de Pseudo-Hermaphrodisme": "Menstruation (if again the account of the subject is veracious) would confirm this opinion [that she is female]; it has not been given to us, to our great regret, to examine Marie C . . . at the moment of a menstrual period" (p. 143).

123. See Ornella Moscucci, "Hermaphroditism and Sex Difference: The Construction of Gender in Victorian England," in Marina Benjamin, ed., *Science and Sensibility: Gender and Scientific Inquiry, 1780–1945* (Cambridge, Mass.: Basil Blackwell, 1991), pp. 174–199.

124. See Rushton Parker, "A Menstruating Man," p. 272.

125. On this question, see, for example, Stonham, "Case of Perfect Uterus Masculinus," p. 226. In France, one author insisted that apparent substitute menstrual signs and even genuine hemorrhages did "not [necessarily] imply the existence of an ovary" because "individuals of both sexes sometimes have hemorrhoidal eruptions when they enter the age of puberty. Sometimes we can establish at the same time [as the eruptions] inflammatory distentions of the breasts in adolescents of both sexes" (Magitot, "Sur un nouveau cas," p. 493).

126. Barnes, "Report of a Living Specimen," p. 210. For the history and anthropology of male menstruation, see James L. Brain, "Male Menstruation in History and Anthropology," *Journal of Psychohistory,* 15 (1988): 311–323.

127. Lutaud, "De L'Hermaphrodisme," p. 392.

128. Cameron, "Notes on a Case of Hermaphrodism," p. 349. In Pozzi, "Sur un pseudo-hermaphrodite," we read: "The breasts are very large and absolutely feminine. The mammae are well developed, and by palpation one feels the lobules of the mammary gland" (p. 135). See also Wood, "The Pelvis and Genital Organs": "The mammae were womanlike and fairly developed, with somewhat prominent nipples. The left mamma weighed when separated [at autopsy] two and three-quarter ounces, and the right two and one-quarter ounces" (p. 169). The breasts were listed among "the characters of femininity" by Polaillon, "Un Cas d'hermaphrodisme": "The anterior part of his/her chest carried two well-formed nipples, surmounting two slightly projecting mammae, very marked projections uncommon in man" (p. 163). On size of breasts, see Samuel Pozzi, "Homme hypospade (pseudo-hermaphrodite) considéré depuis vingt-huit ans comme femme," *Bulletin de la société de médecine légale de France,* 8 (1885): 307–317, p. 316. For the flat nature of the masculine chest, see Barnes, "Report of a Living Specimen," pp. 206, 212; Croom, "Two Cases," p. 102; and Maude, "A Case of Pseudo-Hermaphroditism," p. 429.

129. See, for example, Gérin-Roze, "Un Cas d'hermaphrodisme faux," p. 370.

130. Clark, "A Case of Spurious Hermaphrodism," p. 718.

131. Dandois, "Un Exemple d'erreur de sexe par suite d'hermaphrodisme apparent," *Revue médicale,* 5 (1886): 49–52, p. 50. On undeveloped breasts, see also William C. Hills, "A Case of Hermaphroditism," *Lancet,* 1 (January 25, 1873): 129–130, esp. p. 129.

132. See Guermonprez, "Un erreur de sexe," p. 365; see also Émile Laurent, *Les Bisexués: Gynécomastes et hermaphrodites* (Paris: Georges Carré, 1894).

133. Pozzi, "Homme hypospade," p. 316, original emphasis.

134. Ibid., p. 308. For more on gynecomastia, see also Martel, "Hypertrophie du sein gauche chez un homme atteint de varicocèle du même côté," *Archives de médecine navale,* 60 (1893): 152–154, esp. p. 154; Hutchinson, "Gynaecomazia and Other Aberrations"; and Jonathan Hutchinson, "On Gynaecomazia in Reference to the Development of the Testicles; Atrophy of the Testes and Overgrowth of the Breasts as a Consequence of Injury to the Head," *Archives of Surgery,* 4 (1893): 76.

135. Francis Ogston, "Select Lectures on Medical Jurisprudence; Lecture III: Doubtful Sex," *Medical Times and Gazette,* 1 (1876): 161–162, 189–191, esp. p. 162.

136. Gérin-Roze, "Un Cas d'hermaphrodisme faux," p. 370.

137. Polaillon, "Un Cas d'hermaphrodisme," p. 163. See Accolas, "Cas de pseudo-hermaphrodisme," p. 141.

138. See, for example, Edis's comments in Barnes, "Report of a Living Specimen," p. 212; see also Pozzi, "Pseudo-hermaphrodite," p. 602.

139. See Phillips, "Four Cases," p. 162.

140. See, for example, Turner, "A Case of Hermaphroditism," p. 1884.

141. For such a chart, see Wood, "The Pelvis and Genital Organs," p. 174.

142. Tait, *Diseases of Women*, p. 22; for similar accounts, see Pozzi and Grattery, "Pseudo-hermaphrodite," p. 308; Pozzi, "Note sur deux nouveaux cas," p. 109; Pozzi, "Sur un pseudo-hermaphrodite," p. 136; and Dandois, "Un Exemple d'erreur de sexe," p. 50.

143. Ogston, "Select Lectures," p. 161, original emphasis.

144. Maude, "A Case of Pseudo-Hermaphroditism," p. 429.

145. Clark, "A Case of Spurious Hermaphroditism," p. 718.

146. Croom, "Two Cases," p. 102.

147. See the remarks of Aveling in Barnes, "Report of a Living Specimen," p. 208; see also Johann Ludwig Casper, *A Handbook of the Practice of Forensic Medicine, Based upon Personal Experience*, vol. 3, "including the bio-thanatology of new-born children, and the first part of the biological division," trans. from the third edition by George William Balfour (London: New Sydenham Society, 1864), p. 253.

148. See, for example, Polaillon, "Un Cas d'hermaphrodisme," pp. 162–163.

149. See Trélat, "De L'Hermaphrodisme féminin (leçon recueillie par M. G. Marchant)," *Journal des connaissances médicales pratiques et de pharmacologie*, 3d ser., 2 (1880): 41–42, 58–59, p. 41; see also Accolas, "Cas de pseudo-hermaphrodisme," p. 142; Pozzi, "Note sur deux nouveaux cas," p. 111.

150. See, for example, Pozzi and Grattery, "Pseudo-hermaphrodite": "This individual, a woman for sixty-nine years, was a man" (p. 308). As a woman "his strength was astonishing, and when one had some hard task to do, one knew to whom to address oneself." Compare Magitot, "Sur un nouveau cas": "The muscular force [of a patient is] calculated at the dynamometer marks 50 kilograms, i.e., it is superior to that of women, of whom the average scarcely exceeds, as one knows, 40 to 50 kilograms" (p. 489).

151. See Polaillon, "Un Cas d'hermaphrodisme," p. 163; Targett, "Two Cases," p. 274.

152. See, for example, Clark, "A Case of Spurious Hermaphroditism," p. 718; see also Magitot, "Sur un nouveau cas," p. 489.

153. See, for example, Pozzi and Grattery, "Pseudo-hermaphrodite": "The cadaver is of a stature notably above the mean in women" (p. 308); Polaillon, "Un Cas d'hermaphrodisme": "N. was in size above average" for women (p. 162); Accolas, "Cas de pseudo-hermaphrodisme": "The stature is above the average" (p. 140); Magitot, "Sur un nouveau cas": "The size of this individual is 1.78 m., i.e., a masculine size" (p. 489); Lutaud, "De L'Hermaphrodisme": "She is of an enlarged size for a woman (1.70 m.)" (p. 392); Pozzi, "Pseudo-hermaphrodite": "The one named Adele H., aged 32, is a stocky individual, of vigorous constitution, measuring in height 1.44 m." (p. 602). French practitioners also remarked much more often than their British counterparts on the shape of a subject's buttocks and thighs. On this, see Gaffe, "Un Cas d'hermaphrodisme": "The buttocks are prominent and enormously rotund"

(p. 65); Pozzi, "Note sur deux nouveaux cas": "The *buttocks* are prominent, fatty" (p. 109); ibid.: "The thighs, fatty, entirely hairless, have rather a feminine aspect" (p. 110); Accolas, "Cas de pseudo-hermaphrodisme": "The pelvis is rather large and recalls by its flair the feminine pelvis. The thighs are very muscular, but present a cushion of cellular tissue which rounds and feminizes them" (pp. 141–142).

154. Accolas, "Cas de pseudo-hermaphrodisme," p. 141.

155. Descoust, "Sur un cas d'hermaphrodisme," p. 87.

156. Lloyd, "Case of Spurious Hermaphroditism."

157. Hutchinson, "Gynaecomazia and Other Aberrations," p. 328.

158. G. F. Blacker and T. W. P. Lawrence, "A Case of True Unilateral Hermaphroditism with Ovotestis Occurring in Man, with a Summary and Criticism of the Recorded Cases of True Hermaphroditism," *Transactions of the Obstetrical Society of London,* 38 (1896): 265–317, p. 290.

159. It should also be noted that only the most commonly noted indicators of sex are reviewed in this chapter, although some of the little-used variations are quite interesting; take for example Fauvelle on feminine respiration (in Pozzi, "Pseudo-hermaphrodite," p. 607): "I have been able to note, in examining this mal-defined personality, that she had, as a feminine characteristic, super-costal respiration, which permits during pregnancy no hindering of that primordial function."

160. On the concept of latent hermaphroditism in Victorian England, see Moscucci, "Hermaphroditism and Sex Difference."

161. Polaillon, "Un Cas d'hermaphrodisme," p. 163.

162. Debout, "Hermaphrodite," *Bulletin de la société de médecine de Rouen,* 2d ser., 4 (1890): 43–44, p. 43.

163. For instance, a female face was supposed "sweet" with "a rather feminine expression" in the eyes (Pozzi, "Pseudo-hermaphrodite," p. 602). See also Rose, "Case of Extreme Hypospadia," for a reference to the "features resembling those of a little girl" (p. 256); and Pozzi, "Présentation," pp. 42–43, on the "masculine aspect" of the face. See also the comments of Aveling in Barnes, "Report of a Living Specimen": "The face was feminine, the throat was decidedly that of a woman, the *pomum Adami* not being at all prominent" (p. 208); also those of Heywood Smith in ibid., "In this case the vault of the skull was eminently male, but the face had assumed a feminine cast" (p. 211). See also Maude, "A Case of Pseudo-Hermaphroditism": "In features she resembles her brothers, who are coarse-featured fellows, rather than her sisters, who are delicately pretty and refined" (p. 429).

164. Guermonprez, "Un Cas d'erreur," p. 339.

165. Pozzi, "Homme hypospade," p. 312.

166. Bantock as quoted in Barnes, "Report of a Living Specimen," p. 210.

167. For example, according to Ogston ("Select Lectures"), in woman "the hair of the head [is] fine and long, and that on the other parts of the body comparatively scanty" (p. 161).

168. See, for example, Maude, "A Case of Pseudo-Hermaphroditism," p. 429.

169. Villemin, "Recueil de faits cliniques: Un cas de pseudo-hermaphrodisme masculin avec interventions chirurgicales multiples," *Gazette des maladies infantiles,* 1, no. 13 (1899): 97–98, p. 97.

170. See Pozzi and Grattery, "Pseudo-hermaphrodite," p. 308; and Tait's comments in Barnes, "Report of a Living Specimen," p. 211; Targett, "Two Cases," p. 273.

171. Barnes, "Report of a Living Specimen," p. 212; see also Pozzi and Grattery, "Pseudo-hermaphrodite," p. 308.

172. Polaillon, "Un Cas d'hermaphrodisme," p. 162

173. Lutaud, "De L'Hermaphrodisme," p. 392. Compare Polaillon, "Sur un cas d'hermaphrodisme," *Bulletin de l'académie de médecine,* 3d ser., 25 (1891): 557–561, in which a "feminine" subject's "skin [was] hairless, fine, lined with moderately thick fatty tissue, giving the trunk and members a thin and rounded form," p. 558. See also Wood, "The Pelvic and Genital Organs": "The skin was soft, smooth, and tolerably fair" (p. 169).

174. See Barnes, "Report of a Living Specimen," p. 205; Pozzi and Grattery, "Pseudo-hermaphrodite," p. 308; and Pozzi, "Pseudo-hermaphrodite," p. 602.

175. Barnes, "Report of a Living Specimen," p. 210.

176. Accolas, "Cas de pseudo-hermaphrodisme," pp. 140–141.

177. See Duhosset, "Sur L'Hermaphrodisme," *Bulletin de la société d'anthropologie de Paris,* 3d ser., 4 (1881): 510–513, esp. pp. 510–511.

178. Pozzi, "Homme hypospade," p. 317, original emphasis.

179. See, for example, Martin, "A Case of Hermaphroditism," p. 36; Polaillon, "Sur un cas d'hermaphrodisme," pp. 558–559.

180. On gender and insanity in Victorian England, see Elaine Showalter, "Victorian Women and Insanity," *Victorian Studies,* 23 (1980): 157–181.

181. Hills, "A Case of Hermaphroditism," p. 130.

182. Lucas-Championniere as quoted in Gaffe, "Un Cas d'hermaphrodisme," p. 67.

183. Routh as quoted in Barnes, "Report of a Living Specimen," p. 208. For more on the history of the concept of race as a mitigating factor in sex traits, see John M. Efron, *Defenders of the Race: Jewish Doctors and Race Science in Fin-de-Siècle Europe* (New Haven: Yale University Press, 1994); Sander L. Gilman, "Black Bodies, White Bodies: Toward an Iconography of Female Sexuality in Late Nineteenth-Century Art, Medicine, and Literature," *Critical Inquiry,* 12 (1985): 204–242; Sander L. Gilman, "The Jewish Body: A 'Footnote,'" *Bulletin of the History of Medicine,* 64 (1990): 588–602. For an account of how beards were used to delineate races as well as sexes, see Londa Schiebinger, *Nature's Body: Gender in the Making of Modern Science* (Boston: Beacon Press, 1993), pp. 120–125.

184. See Wood, "The Pelvis and Genital Organs," p. 169; Pozzi and Grattery, "Pseudo-Hermaphrodite," p. 308; and Duhosset, "Sur L'Hermaphrodisme," p. 511.

185. Casper, *A Handbook of the Practice of Forensic Medicine,* p. 253.

186. For a history of concepts of race and breast shape, see pp. 197–198 in Londa Schiebinger, "Mammals, Primatology, and Sexology," in Roy Porter and Mikuláš

Teich, eds., *Sexual Knowledge, Sexual Science: The History of Attitudes to Sexuality* (Cambridge: Cambridge University Press, 1994), 184–209.

187. See, for instance, Phillips, "Four Cases": "[E]nlargement of the clitoris was, and doubtless now is, very common among Asiatic and African races, especially in Arabia; indeed, Albucasis describes the necessary operation for its removal, thus preventing obstruction to coition and libidinous practices" (pp. 162–163).

188. On this, see, for example, Pozzi, *A Treatise on Gynecology,* p. 454.

189. See Moscucci, "Hermaphroditism and Sex Difference," pp. 184–185.

190. For example, Jonathan Hutchinson regretted in a report that his speculations on the origin of a case of gynecomastia "all depend[ed] on the man's testimony, and he may have been desirous to conceal the real facts" (Hutchinson, "On Gynaecomazia in Reference"). See also Stonham, "Case of Perfect Uterus Masculinus": "In some instances I have only the mothers' statements of the facts, and therefore do not attach much importance to them; in other cases I have had the opportunity of examining the subjects themselves" (p. 222). See also Ogston, "Select Lectures": "[N]o attention ought to be paid to the declarations of the individual, or of his or her relations" (p. 191). Recall also Brouardel's conclusion that "the statements were so exaggerated that I dare not take them completely into account" (Brouardel, "Hermaphrodisme," p. 57).

191. Pozzi, "Pseudo-hermaphrodite," see p. 605.

192. Ibid., p. 607.

4. Hermaphrodites in Love

1. François Jules Octave Guermonprez, "Une Erreur de sexe, avec ses consequences," *Journal des sciences médicales de Lille,* 2 (1892): 337–349, 361–376, p. 337.

2. Ibid., pp. 338–339.

3. Ibid., p. 338.

4. Ibid.

5. Ibid., p. 370.

6. Ibid.; emphasis added.

7. Articulating this general notion, the renowned British surgeon Jonathan Hutchinson declared in 1896 that "roughly speaking, and in a general way" the "special organs of [a] sex are inseparably connected with [that sex's] moral and emotional characteristics" (Jonathan Hutchinson, "A Note on Hermaphrodites," *Archives of Surgery,* 7 (1896): 64–66, p. 65). Compare p. 59 in Trélat, "De L'Hermaphrodisme féminin (leçon recueillie par M. G. Marchant)," *Journal des connaissances médicales pratiques et de pharmacologie,* 3d ser., 2 (1880): 41–42, 58–59: "Louise D. had a uterus and an ovary; the genetic sense thus existed in her" (p. 59).

8. See, for example, Patrick Geddes and J. Arthur Thomson, *The Evolution of Sex* (London: Walter Scott, 1889).

9. Charles Debierre, "Pourquoi dans la nature y a-t-il des mâles et des femelles?" *Semaine médicale,* 14 (1894): 454–456 (translated as "Fecundation, Origin of Sexes, Heredity" in *Medical Week,* 2 [1894]: 495–499).

10. Samuel Pozzi, "Homme hypospade (pseudo-hermaphrodite) considéré depuis vingt-huit ans comme femme," *Bulletin de la société de médecine légale de France,* 8 (1885): 307–317; quotations from pp. 312, 313, 316.

11. A. E. Macgillivray, "Operation on an Hermaphrodite; Castration," *Veterinarian* (London), 4th ser. 70 (1897): 580–582, p. 581.

12. Ibid., p. 582.

13. See Émile Laurent, *Les Bisexués: Gynécomastes et hermaphrodites* (Paris: Georges Carré, 1894), p. 205.

14. Adrien Pozzi and P. Grattery, "Pseudo-hermaphrodite (hypospadias périnéal)," *Le Progrès médical,* 2d ser., 5 (1887): 308–312, p. 308, original emphasis. Adrien Pozzi was the brother of the surgeon Samuel Pozzi.

15. Guermonprez, "Une Erreur de sexe," p. 376; original emphasis. Guermonprez did clarify in a footnote that if the hernia had been strangulated, he would have aided the patient. "There is a general rule which rallies all surgeons: an operation destined to save a life is *never* refused" (p. 375, n. 2; original emphasis).

16. John H. Webster, "Case of Hermaphrodism," *British Medical Journal,* 1 (February 10, 1866): 145–146, p. 146. For another example of taking sexual desires as diagnostically significant, see J. Halliday Croom, "Two Cases of Mistaken Sex in Adult Life," *Transactions of the Edinburgh Obstetrical Society,* 24 (1898–99): 102–104, p. 104n.

17. Franz Neugebauer, "Hermaphrodism in the Daily Practice of Medicine; Being Information upon Hermaphrodism Indispensable to the Practitioner," *British Gynaecological Journal,* 29 (1903): 226–263, p. 245.

18. Lawson Tait, *Diseases of Women* (New York: William Wood & Company, 1879), p. 22.

19. Neugebauer, "Hermaphrodism in the Daily Practice," p. 244.

20. Ibid., pp. 244–245.

21. Laurent, *Les Bisexués,* p. 217. Clearly medical men often had the "scandalous" sex life of Alexina/Abel Barbin in mind when they imagined the possible exploits of hermaphrodites. On the issue of wolfish sheep and raising doubtful children male, see also Th. Tuffier and A. Lapointe, "L'Hermaphrodisme. Ses variétés et ses conséquences pour la pratique médicale (d'après un cas personnel)," *Revue de gynécologie et de chirurgie abdominale,* 17 (1911): 209–268; note page 257 where the authors recommend raising doubtful children as males "at least provisionally" because "the presence of a mistaken boy among girls is to be much more feared than the presence of a deformed girl among boys, given the value of feminine virginity from the moral and social point of view. Androgynes [meaning masked males] sometimes have a taste for women and one can imagine the danger which can result when one of these subjects lives by error with feminine subjects." Tuffier and Lapointe also believed that

such a sex-assignment practice would, by sheer statistics, wind up with at least three-quarters of all doubtful children being raised according to the right sex and would also be "a good means for diminishing the number of homosexual unions."

22. Pierre Garnier, "Du Pseudo-hermaphrodisme comme impédiment médico-légale a la déclaration du sexe dans l'acte de naissance," *Annales d'hygiene publique et de médecine légale,* 3d ser., 14 (1885): 285–293, p. 289.

23. Ibid.

24. Ironically, at about the same time, the fear arose that single-sex boarding schools were driving children toward homosexuality. See, for example, the discussion in Havelock Ellis, *Studies in the Psychology of Sex: Sexual Inversion* (Watford, London, and Leipzig: University Press, 1900), p. 37.

25. See the comments on accidental homosexual unions in Tuffier and Lapointe, "L'Hermaphrodisme," p. 257.

26. Laurent, *Les Bisexués,* p. 110.

27. Ibid., p. 181.

28. Thomas H. Wakley and Thomas Wakley, Jr., eds., "Social Leprosy," *Lancet,* 1 (February 2, 1901), p. 346.

29. Robert A. Nye, "The History of Sexuality in Context: National Sexological Traditions," *Science in Context,* 4 (1991): 387–406, p. 393.

30. Xavier Delore, "Des Étapes de l'hermaphrodisme," *L'Écho médicale de Lyon,* 4 (1899): 193–205, 225–232, p. 229.

31. Pozzi, "Homme hypospade," p. 308.

32. Albert Leblond, "Du Pseudo-hermaphrodisme comme impédiment médico-légal a la déclaration du sexe dans l'acte de naissance," *Annales de gynécologie* 24 (1885): 25–35, p. 26.

33. On "marriages made between persons of the same sex," see especially Franz Neugebauer, "Cinquante cas de mariages conclus entre des personnes du même sexe, avec plusieurs procès de divorce par suite d'erreur de sexe," *Revue de gynécologie et de chirurgie abdominale,* 3 (1899): 195–210; Franz Neugebauer, "Une Nouvelle série de 29 observations d'erreur de sexe," *Revue de gynécologie et de chirurgie abdominale,* 4 (1900): 133–174; Franz Neugebauer, "Quarante-quatre erreurs de sexe révélées par l'opération. Soixante-douze opérations chirurgicales d'urgence, de complaisance ou de complicité pratiquées chez des pseudo-hermaphrodites et personnes de sexe douteux," *Revue de gynécologie et de chirurgie abdominale,* 4 (1900): 457–518. "Error of nomenclature" is a phrase taken from Tait, *Diseases of Women,* p. 21.

34. The difference in publicity levels might explain why the French lamented that hermaphrodites "literally [ran] about on the streets," and warned that any infertile couple might be victims of mistaken sex (Garnier, "Du Pseudo-hermaphrodisme," p. 287), suspicions the British did not seem to share.

35. It appears that in France, children and adolescents were more likely than adults to have their sex role suddenly reversed, the assumption being that the emotional clay was not yet so hard that it would crack under the strain of remolding. Thus, for

instance, in March of 1899 a surgeon at the Hospital for Sick Children in Paris reported to the *Gazette des maladies infantiles* the sex "rectification" of an apparent-girl who had originally been brought in by her mother to have the child's hernial bandage replaced. Upon examination, "it was given to us to understand that it was not one, but several operations which would be necessary . . . that it was not the bandage which needed replacing but the civil status of the subject. . . . This large girl in long skirts and opulent hair was a boy" (Villemin, "Recueil de faits cliniques: Un cas de pseudo-hermaphrodisme masculin avec interventions chirurgicales multiples," *Gazette des maladies infantiles*, 1, no. 13 [1899]: 97–98, p. 97). A series of surgical procedures alleviated the awkward nature of his hypospadias—awkward for a male who was supposed to urinate standing up—and the child was renamed and newly costumed as a boy. According to the attending surgeon, "the child has rapidly habituated to his . . . masculine vestments and to his new social condition" (Villemin, "Un cas," p. 98). Both French and British medical men ideally hoped to diagnose "true sex" as early as possible, to avoid all the problems that could ensue from mistaken sex. No one doubted that it was "of the highest importance for a young individual to have a [true] sex as soon as possible, for it determines the direction of the entire life and indicates rights and duties" (Xavier Delore, "De L'Hermaphrodisme dans l'histoire ancienne et dans la chirurgie moderne," *Journal des sciences médicales de Lille*, 2 [1899]: 63–70, p. 68).

36. Alcide Benoist, "Rapport sur un cas d'hermaphrodisme," *Annales d'hygiene publique et de médecine légale*, 3d ser., 16 (1886): 84–87, p. 85.

37. Recall for instance, the two adult sisters in Scotland who were informed that they were both men. The siblings had their sexual identities quietly changed and were dispatched to pursue manly occupations in the colonies where no one would know they had once been women. See Chapter 3 and Croom, "Two Cases."

38. George Buchanan, "Hermaphrodite, Aged 9, in Whom Two Testicles Were Excised from the Labia Majora," *Glasgow Medical Journal*, 23 (1885): 213–217, p. 214.

39. Ibid., p. 215.

40. This contradicts Thomas Laqueur's claim that, while ovariotomies were not uncommon during this period, testicles were "sacrosanct," surgically speaking (Thomas Laqueur, *Making Sex: Body and Gender from the Greeks to Freud* [Cambridge: Harvard University Press, 1990], p. 177), although these cases (in which "women" had "testicles") were assumed to be rather extraordinary.

41. Andrew Clark, "A Case of Spurious Hermaphroditism (hypospadias and undescended testes in a subject who had been brought up as a female and been married for sixteen years)," *Lancet*, 1 (March 12, 1898): 718–719, p. 719.

42. G. R. Green, "A Case of Hypospadia in a Patient, Aged 24, Who Had Always Passed as a Woman," *Quarterly Medical Journal* (Sheffield), 6 (1898): 130–132, p. 130.

43. Ibid., p. 132.

44. Ibid.

45. Delore, "Des Étapes," p. 229.

46. Raoul Blondel, "Observation de pseudo-hermaphrodisme.—Un homme marié à un homme," *Journal de médecine de Paris,* 2d ser., 11 (1899): 75–77, p. 75.

47. For a similar case in which a potential wife expressed interest in a hermaphroditic "man" specifically because he was probably sterile, see Gaffe, "Un Cas d'hermaphrodisme," *Journal de médecine et de chirurgie pratiques,* 56 (1885): 65–67.

48. Blondel, "Observation," p. 76.

49. Ibid., p. 77.

50. Neugebauer, "Cinquante cas," p. 198.

51. Ibid.

52. Blondel, "Observation," p. 77.

53. Debout, "Hermaphrodite," *Bulletin de la société de médecine de Rouen,* 2d ser., 4 (1890): 43–44, p. 44.

54. See also the case of Henriette Williams in which a French doctor tried to steer that "mistaken" woman away from "same"-sex relations when, without specifying why, he simply told her that it would be improper for her to enter marriage as a woman and that she should avoid relations with men. (Lutaud, "De L'Hermaphrodisme au point de vue médico-légal.—Nouvelle observation.—Henriette Williams," *Journal de médecine de Paris,* [1885]: 386–396.) Interestingly, the author of this report left this medical man—whom the author admonished—anonymous, referring to him only as "a distinguished doctor in Paris, Dr. L.D."

55. Delore, "Des Étapes," p. 229.

56. Ibid.

57. Ibid.

58. Delore, "De L'Hermaphrodisme," p. 70.

59. Tuffier and Lapointe, "L'Hermaphrodisme," p. 266; original emphasis. Compare the remarks of the medico-legal expert Auguste Tardieu in his 1874 discussion of Barbin's life and death: "[S]cience and the law . . . [are] obliged to recognize error and to recognize true sex," that is, as determined by the gonads; Tardieu as quoted on page 123 of Herculine Barbin, *Herculine Barbin. Being the Recently Discovered Memoirs of a Nineteenth-Century French Hermaphrodite,* intro. Michel Foucault, trans. Richard McDougall (New York: Pantheon, 1980). See also page 301 in Jean Tapie, "Un Cas d'erreur sur le sexe. Malformation des organes génitaux externes; pathogénie de ces vices de conformation," *Revue médicale de Toulouse,* 22 (1888): 301–313.

60. In fact, one reason Blondel maintained some doubt as to Madame X's "true sex" was that her "tastes and penchants were clearly feminine, and at every moment the subject indicated no inclination for women" (Blondel, "Observation," p. 77).

61. Laurent, *Les Bisexués,* p. 206.

62. Louis Blanc, *Les Anomalies chez l'homme et les mammifères* (Paris: J.-B. Baillière et Fils, 1893), p. 209. Still, Blanc assured his readers, usually the "true" sex (and sexuality) would emerge, and troubles would end: "In more than one case, the individual, whose sex has been the subject of error, presented during a part of his life

the tendencies of his supposed sex, then recouped spontaneously *his natural instincts*" (p. 209, original emphasis). Blanc offered as an example a classic story of the hermaphrodite made good: "Worbe reported the history of a hermaphrodite born in 1775 in the area of Dreux, and baptized under the name Marie-Jeanne. Raised a girl, this subject for a long time wore the dress of the young girl villagers; but, toward the age of 20 years, the masculine tendencies made themselves known: Marie-Jeanne frequented the cabarets, got drunk often, smoked a pipe, hunted, [and] was pleased to lead and attend horses. Later she took on the clothes of her true sex, and married" (pp. 209–210).

63. Michel Foucault, *The History of Sexuality,* vol. 1: *An Introduction,* trans. Robert Hurley (New York: Random House, 1978), p. 43; see also Arnold Davidson, "Sex and the Emergence of Sexuality," in Edward Stein, ed., *Forms of Desire: Sexual Orientation and the Social Constructionist Controversy* (New York: Routledge, 1990), pp. 89–132; and Jonathan Ned Katz, *The Invention of Heterosexuality* (New York: Dutton, 1995).

64. And by implication the bisexual, who is commonly seen as a combination of a heterosexual and a homosexual.

65. See Vernon A. Rosario, "Inversion's Histories/History's Inversions: Novelizing Fin-de-Siècle Homosexuality," in Vernon A. Rosario, ed., *Science and Homosexualities* (New York: Routledge, 1997), pp. 89–107, see p. 91.

66. See Valentin Magnan and Samuel Pozzi, "Inversion du sens génital chez un pseudo-hermaphrodite féminin. Sarcome de l'ovaire gauche opéré avec succès," *Bulletin de l'académie de médecine,* 3d ser., 65 (1911): 223–259.

67. Ibid., p. 228.

68. Ibid., p. 229.

69. Pozzi did, however, suggest generally that if "the seminal glands (in the immense majority of cases [of hermaphroditism], testicles) are . . . ectopic and sensibly atrophied, or, to say it better, incompletely developed," then "these individuals, who present sexual glands reduced to a minimum, ought to have very weak reflex genital actions" (ibid., p. 234).

70. Ibid., p. 232.

71. Ibid., p. 233.

72. Ibid., p. 234; original emphasis.

73. Ibid., p. 234; original emphases.

74. Indeed, even though they knew very well that "women [did] not interest [L.S.] from the genital point of view," the medical men summarizing the case of L.S. maintained a sneaking suspicion that L.S. would "sooner or later have relations, not only with normal men as she does today, but also with [male] inverted homosexuals and with women" (Tuffier and Lapointe, "L'Hermaphrodisme," p. 219). That is, she would slowly come around.

75. Bert Hansen has noted that British physicians remained "rather reticent on this subject through the 1890s"; Hansen, "American Physicians' 'Discovery' of Homo-

sexuals, 1880–1900: A New Diagnosis in a Changing Society," in Charles E. Rosenberg and Janet Golden, eds., *Framing Disease* (New Brunswick, N.J.: Rutgers University Press, 1992), p. 105. See also Ellis, *Sexual Inversion,* pp. xi–xii.

76. Interestingly, no doubt was thrown by medical men on the sex or sexuality of "homosexual" hermaphrodites' sex partners. Logically, if a true woman would desire a true man only, women partners should not have been "deceived" into loving female hermaphrodites who looked like men; similarly, men partners should not have been "deceived" into loving male hermaphrodites who looked like women. (I am grateful to Vernon Rosario for pointing this out.)

77. Hutchinson, "A Note on Hermaphrodites," p. 66.

78. See, for example, William Lee Howard, "Psychical Hermaphroditism. A Few Notes on Sexual Perversion, with Two Clinical Cases of Sexual Inversion," *Alienist and Neurologist,* 18 (1897): 111–118.

79. See, for example, p. 170 in Havelock Ellis, *Studies in the Psychology of Sex: Sexual Inversion,* 2d ed. (Philadelphia: F. A. Davis Co., 1908), p. 170; and Laurent, *Les Bisexués,* p. 20.

80. On this, see Rosario, "Inversion's Histories," p. 90.

81. See, for example, Appendix A, "A Categoric Personal Analysis for the Reader," in Xavier Mayne (pseudonym for Edward Irenaeus Prime Stevenson), *The Intersexes: A History of Similisexualism as a Problem in Social Life* (originally privately printed, n.d.; preface dated 1908; reprinted, New York: Arno Press, 1975); and see the "Questionnaire on Homosexuality" by Earl Lind reproduced in Hansen, "American Physicians' 'Discovery,'" pp. 112–113; and Rosario, "Inversion's Histories," p. 97.

82. Richard von Krafft-Ebing, *Psychopathia Sexualis, with Special Reference to Contrary Sexual Instinct,* authorized translation of the seventh enlarged and revised German edition, trans. Charles Gilbert Chaddock (Philadelphia: F. A. Davis Company, 1921), p. 222.

83. Nye, "The History of Sexuality," p. 399.

84. See Rosario, "Inversion's Histories," p. 92.

85. Ellis, *Sexual Inversion* (1908), p. 171; emphasis added.

86. Krafft-Ebing, *Psychopathia Sexualis,* p. 304.

87. Mayne, *The Intersexes: A History of Similisexualism,* p. 72.

88. Ibid., p. 124.

89. Lillian Faderman, "The Morbidification of Love between Women by Nineteenth-Century Sexologists," *Journal of Homosexuality,* 41 (1978): 73–90, p. 81.

90. The usage of the term "psycho-sexual hermaphrodite" to refer to homosexuals was especially common among American medical men; see, for example, Howard, "Psychical Hermaphroditism," and C. W. Allen, "Report of a Case of Psycho-Sexual Hermaphroditism," *Medical Record* (New York), 51 (1897): 653–655. For use of the term to refer to what we call bisexuals, see Ellis, *Sexual Inversion* (1900), p. 73; see also Krafft-Ebing, *Psychopathia Sexualis,* p. 230.

91. See Ellis, *Sexual Inversion* (1900), p. 73.

92. Hubert C. Kennedy, "The 'Third Sex' Theory of Karl Heinrich Ulrichs," *Journal of Homosexuality,* 6 (1980/1981): 103–111, p. 103.

93. Vern L. Bullough, *Science in the Bedroom: A History of Sex Research* (New York: Basic Books, 1994), p. 34.

94. The nomenclature of Uranian, Uraniad, and so forth was developed by Ulrichs. For an elaboration of it and his thinking, see Bullough, *Science in the Bedroom,* pp. 35–37, and Kennedy, "The 'Third Sex' Theory."

95. Mayne, *The Intersexes,* p. x.

96. Ibid., p. 14.

97. Ibid., p. 16.

98. Rosario, "Inversion's Histories," p. 97.

99. On this notion, see Chapter 2 and Stephen Jay Gould, *Ontogeny and Phylogeny* (Cambridge: Harvard University Press, 1977), pp. 49–52.

100. On this assumption, see Nye, "The History of Sexuality," pp. 395, 399; and see Robert A. Nye, "Sex Difference and Male Homosexuality in French Medical Discourse, 1830–1930," *Bulletin of the History of Medicine,* 63 (1989): 32–51, p. 36.

101. Laurent, *Les Bisexués,* p. 138.

102. Ellis, *Sexual Inversion* (1908), p. 188.

103. Ibid., p. 186.

104. Nye, "The History of Sexuality," p. 391.

105. Delore, "De L'Hermaphrodisme," p. 70.

106. Delore, "Des Étapes," p. 226.

5. The Age of Gonads

1. See Londa Schiebinger, *Nature's Body: Gender in the Making of Modern Science* (Boston: Beacon Press, 1993), chap. 2.

2. The four kingdoms (or *embranchemens*) were arranged as follows: I. *Hémitéries,* including simple varieties (like color variations) and "vices of conformation," that is, relatively minor anomalies which might nonetheless disrupt normal functioning; II. *Hétérotaxies,* including anomalies which appear to be serious anatomically but which do not disrupt normal functioning; III. Hermaphroditisms, or any mixture of ordinarily distinct male and female traits; and IV. Monstrosities, or the most complex, serious anomalies which render normal function impossible or difficult. Taken from Isidore Geoffroy Saint-Hilaire, *Histoire générale et particulière des anomalies de l'organization chez l'homme et les animaux . . . ou Traité de tératologie* (Paris: J.-B. Baillière, 1832–36), vol. 1, pp. 32–36.

3. See Isidore Geoffroy Saint-Hilaire, "Recherches anatomiques et physiologiques sur l'hermaphrodisme anormal chez l'homme et chez les animaux," presented to the Académie des Sciences on February 4, 1833. The system was shortly thereafter published in more detail in the second volume of his *Traité de tératologie.* For a summary review, see Alice Domurat Dreger, "'Nature Is One Whole': Isidore Geof-

froy Saint-Hilaire's *Traité de tératologie*" (M.A. thesis, Indiana University, 1993), pp. 106–122.

4. Based on Geoffroy Saint-Hilaire, *Traité de tératologie,* vol. 2, p. 36.

5. Mixed hermaphroditism (Class I, Order 4) was further subdivided into four types as follows: (a) superposed hermaphrodism, male and female organs being superposed; (b) semilateral hermaphrodism, organs of one side all of the same sex, those of the other are male and female; (c) lateral hermaphrodism, right-side organs of one sex, left-side of the other; (d) crossed hermaphrodism, the profound organs of the right side and the middle of the left side being of one sex, the others of the other sex.

6. This division into "with excess in the number of parts" and "without excess" was typical of Geoffroy's teratological taxonomies. Geoffroy secondarily arranged kinds of hermaphroditisms from the most simple to the most complex. An example of the most simple would be a hermaphroditism in which the sexual apparatus was "essentially female," that is, female in the profound parts, but included a few "male" or "hermaphroditic" external elements in place of the typical female ones. The most complex hermaphroditism was not termed "true" by Geoffroy. Instead the most complex hermaphroditism was "perfect bisexual hermaphroditism," a condition in which, theoretically by an excess of many body parts, a being would have all the sex segments of both the typical male and the typical female.

7. This is not to imply that Geoffroy found the idea of a profound mixing of the male and female in a human appealing. In fact, while explaining why he thought perfect bisexual hermaphroditism was extremely unlikely ever to be found in humans, Geoffroy articulated his disbelief that male and female sexuality could ever profoundly co-exist in one body: "Although, by the conformation of the double sexual apparatus, it could be materially possible that a hermaphrodite could in turn impregnate and be impregnated, there is strong doubt that its penchants could bring it simultaneously toward the two sexes, or that it could fulfill both functions" (Geoffroy Saint-Hilaire, *Traité de tératologie,* vol. 2, p. 171). Geoffroy also suspected perfect bisexual hermaphroditism would not be found in humans because it would require the coincidence of a huge number of anomalies, and because it would produce such "grave perturbations in the normal connections" that it would probably result in inviability; see vol. 2, p. 172.

8. Geoffroy Saint-Hilaire, *Traité de tératologie,* vol. 2, p. 111.

9. The words "hermaphroditism" and "hermaphrodism" were used interchangeably, but the British typically preferred "hermaphroditism" because it preserved the entire root of the word. Neugebauer and the French tended to use "hermaphrodism" or the French cognate, "hermaphrodisme."

10. James Young Simpson, "Hermaphroditism, or Hermaphrodism," in Robert Bentley Todd, ed., *The Cyclopaedia of Anatomy and Physiology,* vol. 2 (London: Sherwood, Gilbert and Piper, 1839), pp. 684–738, p. 684, emphasis added. (This article was reprinted in Sir W. G. Simpson, ed., *The Works of Sir James Y. Simpson,*

Bart., vol. 2, *Anaesthesia, Hospitalism, Hermaphroditism, and a Proposal to Stamp Out Small-Pox and Other Contagious Diseases* [New York: D. Appleton & Co., 1872].)

11. Reconstructed from ibid., pp. 684–685.

12. For Simpson's treatment of Arsano, see ibid., pp. 703–704; see also an application of this classification in H. G. Rawdon, "Description of a Case of True Hermaphroditism, with Remarks," *Liverpool Medical and Surgical Reports,* 1 (1867): 39–52, p. 47; for expert support of Simpson's system, see Francis Ogston, "Select Lectures on Medical Jurisprudence; Lecture III: Doubtful Sex," *Medical Times and Gazette,* 1 (1876): 161–162, 189–191.

13. Robert Knox, "Hermaphroditism; A Memoir Read to the Royal Society of Edinburgh in 1827 and 1829," *London Medical Gazette,* n.x. (1843): 241–243, 293–300, 447–451, and 472–477, p. 450. For a review of Knox's philosophy of anatomy, see Evelleen Richards, "The 'Moral Anatomy' of Robert Knox: The Interplay between Biological and Social Thought in Victorian Scientific Naturalism," *Journal of the History of Biology,* 22 (1989): 373–436.

14. See K. Codell Carter, "Edwin Klebs' Criteria for Disease Causality," *Medizinhistorisches Journal,* 22 (1987): 80–89, p. 81.

15. This is the classification system for hermaphroditism proposed by Edwin Klebs, *Handbuch der Pathologischen Anatomie* (Berlin: A. Hirschwald, 1876); see vol. 1, p. 718. See also Gabrielle Roos Clog, "Documents historiques sur la notion d'intersexualité," presented for the doctorate in medicine, Faculté de Médecine de Strasbourg, no. 71 (1970), available at the Wellcome Institute Library, London, and see Samuel Pozzi, "Neuf cas personnels de pseudo-hermaphrodisme," *Revue de gynécologie et de chirurgie abdominale* 16 (1911): 269–339, p. 269.

16. Klebs allowed for three subtypes of masculine pseudohermaphroditism: (1) internal (testes; presence of a prostate with "masculine uterus"; masculine externally with some internal rudimentary feminine organs); (2) internal and external (or "complete") masculine pseudohermaphroditism (testes; "masculine uterus" with presence of fallopian tubes; urinary apparatus divided by the uterus; genitalia feminine; mostly feminine externally and internally except for testes); (3) external (testes; somewhat feminine external genitalia; feminine "habit" generally; somewhat or mostly feminine externally; masculine internally). A problem with the system (which Pozzi noted in Pozzi, "Neuf cas personnels," p. 270) was the confusion over to what the adjectives referred. The adjectives "masculine" or "feminine" referred to the nature of the gonads (testicular or ovarian, respectively). The adjectives "internal" or "external" referred to the site of the sexual anomaly (anomalous in comparison with the gonads, which are always assumed to be the norm).

17. Klebs allowed for three types of feminine pseudohermaphroditism: (1) internal (ovaries; internally and externally feminine with some masculine rudimentary internal organs); (2) internal and external feminine (or "complete") (ovaries; a portion of the exterior genital apparatus and of the internal conduits are masculine; mostly masculine externally and internally except for the ovaries); (3) external (ovaries; mostly

external feminine genital apparatus with presence of some external masculine characters; internally feminine but externally somewhat or mostly masculine).

18. Robert Barnes and Fancourt Barnes, *A System of Obstetric Medicine and Surgery, Theoretical and Clinical, for the Student and the Practitioner,* 2 vols. (London: Smith, Elder & Co., 1884–85), vol. 1, p. 531.

19. See, for example, Odin, "Hermaphrodisme bi-sexuel," *Lyon médical,* 16 (1874): 214–218; see also Cunningham, "Hermaphroditism in the Goat," *British Medical Journal,* 1 (June 6, 1885): 1159. The gynecological surgeon Lawson Tait was an early lobbyist for the gonadal definition. In his 1879 textbook *Diseases of Women,* Tait argued that "it is not a philosophical proceeding to say that both sexes are represented unless both a testicle and an ovary are present"; Lawson Tait, *Diseases of Women* (New York: William Wood & Company, 1879), p. 124. Tait did not cite the work of Klebs, but, notably, like Klebs, Tait thought some forms of true hermaphroditism possible in humans. Drawing on his knowledge of embryology, Tait argued that, since the ovary and testicle originate "from the same blastema, and being really the same organ, it is not surprising that occasionally reversions of type should occur, so that an immature testicle should appear on the one side, and an imperfect ovary on the other" (p. 123).

20. Blacker and Lawrence's article apparently constituted the first published introduction of Klebs's classification system to the British medical community.

21. G. F. Blacker and T. W. P. Lawrence, "A Case of True Unilateral Hermaphroditism with Ovotestis Occurring in Man, with a Summary and Criticism of the Recorded Cases of True Hermaphroditism," *Transactions of the Obstetrical Society of London,* 38 (1896): 265–317, p. 266; emphasis added. In the late 1870s and early 1880s, Lawson Tait and Samuel Pozzi both campaigned not only for the idea that the gonads should be the exclusive markers of true sex but also for the notion that, before a conclusive diagnosis was made, the nature of the gonads should be confirmed microscopically. Tait and Pozzi repeatedly insisted that "the only satisfactory test is that of microscopic examination" (p. 124 in Tait, *Diseases of Women;* see also Pozzi's comments on p. 557 of Houel, "Pièces d'hermaphrodites conservées au musée Dupuytren," *Bulletin de la société d'anthropologie de Paris,* 3d ser., 4 [1881]: 554–558), but this criterion was not widely adopted until after Blacker and Lawrence's article.

22. Blacker and Lawrence, "A Case of True Unilateral Hermaphroditism," p. 307.

23. This amendment of Klebs's category allowed Blacker and Lawrence to announce their case as an instance of true hermaphroditism, an important claim when, after their decimation of previous claims, they left only three cases standing. The coauthors were able, using a strict microscopic-anatomical criterion of true sex, to proclaim that "the specimen described in the present paper appears to be the only case of true unilateral hermaphroditism in man recorded up to the present time" (Blacker and Lawrence, "A Case of True Unilateral Hermaphroditism," p. 266). When innovations were introduced to the science of hermaphroditism, it seemed that invariably the new system privileged the findings of the innovator(s). For instance,

Louis Guinard, physiologist at Lyons and author of a treatise on teratology, overtly revised Geoffroy Saint-Hilaire's highest category of hermaphroditism to fit a case of his own into that most rare and interesting—and therefore most important—category. Guinard offered an original case of hermaphroditism in which a goat had ovaries and testes, but not all the organs of both sexes. Geoffroy had indicated that the "perfect bisexual hermaphrodite" had *all* of the organs of both sexes, but Guinard argued that his goat should count as a perfect bisexual hermaphrodite, since it presented "the variety of hermaphroditism which is certainly the most perfect"; Louis Guinard, "Un Cas d'hermaphrodisme parfait bisexuel," *Journal de médecine vétérinaire et de zootechnie,* 3d ser., 15 (1890): 351–353, p. 351. Guinard's definitional revision also marked the trend toward using the sex glands as the exclusive markers of true sex, and therefore of true hermaphroditism.

24. Consequently any case in which an autopsy had not (yet) been performed was eliminated; see, for example, the elimination of Messner's case: "Messner's case suffers from the fact that it is the result of an examination during life only" (Blacker and Lawrence, "A Case of True Unilateral Hermaphroditism," p. 297).

25. For example, see the rejection of Fowler's case in ibid., p. 284.

26. Ibid., p. 286.

27. A very small minority of medical men in Britain continued to diagnose true hermaphroditism in the older, more liberal mode; see, for example, G. R. Turner, "A Case of Hermaphroditism," *Lancet,* 1 (June 30, 1900): 1884–85; and see also Henry Corby, "Removal of a Tumour from a Hermaphrodite," *British Medical Journal,* 2 (September 23, 1905): 710–712. This was probably done in ignorance of the standardization revisions, since they had seen no previous cases and made no issue of the use of the phrase "true hermaphrodite."

28. See, for example, A. Primrose, "A Case of Uterus Masculinus—Tubular Hermaphroditism—in the Male, with Sarcomatous Enlargement of an Undescended Testicle," *Journal of Anatomy and Physiology,* 33 (1898–1899): 64–75, p. 70; and see also Barton Cooke Hirst and George A. Piersol, *Human Monstrosities,* four parts (Edinburgh and London: Young J. Pentland, 1892), p. 82.

29. See J. W. Ballantyne, "Hermaphroditism," in Chalmers Watson, ed., *Encyclopaedia Medica,* vol. 4 (Edinburgh: William Green and Sons, 1900), p. 491.

30. Compare Samuel Pozzi, "De L'Hermaphrodisme," *Gazette hebdomadaire de médecine et de chirurgie,* 2d ser., 27 (1890): 351–355; on page 354 we read: "One can accord a real importance to the fact [the fact of true hermaphroditism] only following the autopsy."

31. Jonathan Hutchinson, "A Note on Hermaphrodites," *Archives of Surgery,* 7 (1896): 64–66, p. 65.

32. Pierre Garnier, "Du Pseudo-hermaphrodisme comme impédiment médico-légale a la déclaration du sexe dans l'acte de naissance," *Annales d'hygiene publique et de médecine légale,* 3d ser., 14 (1885): 285–293, p. 287. Compare Trélat, "De L'Hermaphrodisme féminin (leçon recueillie par M. G. Marchant)," *Journal des*

connaissances médicales pratiques et de pharmacologie, 3d ser., 2 (1880): 41–42, 58–59: "the hermaphrodite is an individual with bisexual genital organs but in whom at least *one* of the organs is always insufficient, defective" (p. 41; original emphasis).

33. See Thomas Neville Bonner, *Becoming a Physician: Medical Education in Britain, France, Germany, and the United States, 1750–1945* (New York: Oxford University Press, 1995), p. 279.

34. See, for example, the embryological descriptions in Oscar Hertwig, *Text-Book of the Embryology of Man and Mammals,* trans. from the third edition by Edward L. Mark (London: Swan Sonnenschein & Company, 1892), p. 374, and in Charles Sedgwick Minot, *Human Embryology* (New York: William Wood, 1892), pp. 492–496.

35. Minot, *Human Embryology,* p. 84; original emphasis.

36. Ibid., p. 84.

37. Jean Tapie, "Un Cas d'erreur sur le sexe. Malformation des organes génitaux externes; pathogénie de ces vices de conformation," *Revue médicale de Toulouse* 22 (1888): 301–313, p. 305.

38. See, for example, Patrick Geddes and J. Arthur Thomson, *The Evolution of Sex* (London: Walter Scott, 1889); and Charles Debierre, "Pourquoi dans la nature y a-t-il des mâles et des femelles?" *Semaine médicale,* 14 (1894): 454–456 (translated as "Fecundation, Origin of Sexes, Heredity" in *Medical Week,* 2 [1894]: 495–499).

39. At an 1881 discussion on hermaphroditism at the Paris Société d'Anthropologie, for example, the surgeon Samuel Pozzi insisted that "that which determines [the sex] is the *presence* of either testicles or ovaries" (quoted in Duhosset, "Sur L'Hermaphrodisme," *Bulletin de la société d'anthropologie de Paris,* 3d ser., 4 [1881]: 510–513, p. 513; emphasis added). Pozzi was adamant that a gonadal definition of sex was the only one that made sense, and in 1881 he managed to convince the Paris Society of Surgery to adopt as its doctrine of true hermaphroditism the strict criteria of the presence of the "noble organs," that is, the gonads, of both sexes (this according to Jouin, "Hermaphrodisme vrai et pseudo-hermaphrodisme," *Bulletins et mémoires de la société obstetricale et gynécologique de Paris,* [1891]: 190–197, pp. 190–191). Throughout his career, Pozzi remained convinced that "properly speaking there are in the human species no true hermaphrodites" (Samuel Pozzi, "Présentation d'un pseudo-hermaphrodite male," *Comptes rendus des séances de la société de biologie,* 8th ser., 1, pt. 2 [1884]: 42–45, p. 44).

40. See Merriley Elaine Borell, "Organotherapy and the Emergence of Reproductive Endocrinology," *Journal of the History of Biology,* 18 (1985): 1–30. See also Merriley Borell, "Organotherapy, British Physiology, and Discovery of the Internal Secretions," *Journal of the History of Biology,* 9 (1976): 235–268; and Nelly Oudshoorn, *Beyond the Natural Body: An Archeology of Sex Hormones* (New York: Routledge, 1994), esp. chap. 2.

41. Blacker and Lawrence, "A Case of True Unilateral Hermaphroditism," p. 266. In 1905 Lawrence reached out to internal secretion theories for a logic to why one would require only both kinds of tissue, rather than both organs, for true hermaph-

roditism: "[M]ere fractions appear to exert an influence hardly inferior to that of the whole gland" (T. W. P. Lawrence, "True Hermaphroditism in the Human Subject," *Transactions of the Pathological Society of London,* 57 [1906]: 21–44, p. 24). (Endocrinological considerations of sex are discussed in more detail below.)

42. The German Ahfeld, for instance, defined the true hermaphrodite as "an individual with functionally active glands of both sexes, provided with excretory ducts," that is, one potentially able to impregnate him/herself. But if such a reproductive-capability criterion had been used in parallel for spurious hermaphrodites and nonhermaphrodites, many of them would have appeared sexless. (On Ahfeld, see Hirst and Piersol, *Human Monstrosities,* p. 75.)

43. "Feminism" in Henry Power and Leonard W. Sedgwick, *The New Sydenham Society's Lexicon of Medicine and the Allied Sciences,* 5 vols. (London: New Sydenham Society, 1881–1889).

44. Blacker and Lawrence, "A Case of True Unilateral Hermaphroditism," p. 266.

45. See, for example, Pozzi, "Neuf cas personnels," p. 273; Dawson Williams and Charles Louis Taylor, eds., "Hermaphroditism," *British Medical Journal,* 2 (September 23, 1911): 694–695, p. 695.

46. Pozzi, "Neuf cas personnels," p. 270; see also Th. Tuffier and A. Lapointe, "L'Hermaphrodisme. Ses variétés et ses conséquences pour la pratique médicale (d'après un cas personnel)," *Revue de gynécologie et de chirurgie abdominale,* 17 (1911): 209–268, p. 226.

47. Pozzi, "Neuf cas personnels," pp. 270–271; emphasis added.

48. For instance, a 1911 report in the *British Medical Journal,* which updated British medical men on French studies of hermaphroditism, announced that "the structure of the genital gland settles the sex [of pseudohermaphrodites], yet at the same time does not control the anatomical development of the genital tract [in hermaphrodites]. Still less does it settle secondary characteristics" (Williams and Taylor, "Hermaphroditism," p. 694).

49. Pozzi, "Neuf cas personnels," p. 269.

50. See ibid., p. 270.

51. The complete system was presented in ibid., p. 274. For preliminary versions, see Pozzi, "De L'Hermaphrodisme," p. 352; Samuel Pozzi, "Sur un pseudo-hermaphrodite androgynoïde; prétendue femme ayant de chaque côté un testicule, un épididyme (ou trompe?) kystique et une corne utérine rudimentaire à gauche formant hernie dans le canal inguinal. Cure radicale. Examen microscopique," *Bulletin de l'académie nationale de médecine,* 3d ser., 36 (1896): 132–145; Samuel Pozzi, "Pseudo-hermaphrodite androgynoïde," *Presse médicale,* no. 64 (August 6, 1896): cccxiii–cccxv.

52. Pozzi hyphenated the words to enable easy pronunciation. The following was the classification system proposed by Pozzi and presented to the Académie de Médecine on February 21, 1911 (see Pozzi, "Neuf cas personnels," p. 274):

I. Andro-gynoides (true men [have testes] with some feminine traits):

Regular: normal (feminine) morphology of the vulva with histologically demonstrated testicles.

Irregular: Vulviform perineo-scrotal hypospadias; clitoridean penis, with testicles histologically demonstrated.

II. Gyn-androids (true women [that is, have ovaries] with some masculine traits):
Regular: Normal (masculine) morphology of the external genital organs; canaliculi penis; absence of the vulva; perineal raphe. With ovaries histologically demonstrated.

Irregular: Aplastic vulvo-vagina; peniform clitoris. With ovaries histologically demonstrated.

53. David Berry Hart, "On the Atypical Male and Female *Sex-Ensemble* (So-Called Hermaphroditism and Pseudohermaphroditism)," *Edinburgh Medical Journal*, n.s., 13 (1914): 295–316, p. 295.

54. Hart was the first to study the structural anatomy of the female pelvis by using frozen sectioning according to J. W. Ballantyne, "Obituary [of] David Berry Hart, M.D., F.R.C.P.E.," *Edinburgh Medical Journal*, n.s., 25 (1920): 122–129, p. 122. Hart also became actively involved in the debate over the nature of the free-martin (see Chapter 2); and he tried to explain normal sex differentiation using the concept of Mendelian unit characters (Hart, "On the Atypical," pp. 298ff.).

55. Hart, "On the Atypical," p. 295.

56. Ibid., p. 296.

57. Ibid.

58. Ibid.

59. See, for example, the cases of Henriette Williams and Madame X. of Angers.

60. Jordan Lloyd, "Case of Spurious Hermaphroditism," *Illustrated Medical News*, 2 (February 2, 1889): 103–104, p. 103.

61. William Blair Bell, "Hermaphroditism," *Liverpool Medico-Chirurgical Journal*, 35 (1915): 272–292, p. 291.

62. Ibid., p. 279.

63. Ibid., p. 279–280.

64. Ibid., p. 280.

65. Ibid., p. 281.

66. Ibid.

67. Ibid.

68. Ibid., p. 291; see also Tuffier and Lapointe, "L'Hermaphrodisme," p. 226.

69. Blair Bell, "Hermaphroditism," p. 275. No one imagined that the latter condition, called by Tuffier and Lapointe ("L'Hermaphrodisme") "total hermaphroditism," was ever realized in humans.

70. Tuffier and Lapointe, "L'Hermaphrodisme," p. 256.

71. Blair Bell, "Hermaphroditism," p. 287.

72. Ibid., p. 291.

73. Ibid., emphasis added.

74. Ibid., p. 289.

75. Ibid., pp. 287–288.

76. Ibid., p. 288.

77. Ibid., pp. 289–290.

78. Ibid., p. 277; emphases added.

79. Ibid., p. 292.

80. Ibid.

Epilogue

1. Donald Bateman, "The Good Bleed Guide: A Patient's Story," *Social History of Medicine,* 7 (1994): 115–133, p. 115.

2. For a view of objectivity that incorporates and even requires this kind of "listening," see Lisa M. Heldke and Stephen H. Kellert, "Objectivity as Responsibility," *Metaphilosophy,* 26 (1995): 360–378.

3. Joseph Campbell, *The Hero with a Thousand Faces* (Princeton: Princeton University Press, 1949, 1968).

4. Arthur W. Frank, *The Wounded Storyteller: Body, Illness, and Ethics* (Chicago: University of Chicago Press, 1995), p. 158. I am grateful to Albert Arias for suggesting that I read Frank's work and for providing me his reading of how it relates to intersexuals' stories.

5. Ibid., p. 163. Although Frank is concerned chiefly with illness narratives, he recognizes that they have much in common with many other suffering narratives, for instance, the stories told by and about Holocaust survivors. Hermaphroditism is not an illness in the sense that, in itself, it is not an anatomically degenerative state—although its correlation with certain disease states is elaborated below. Intersexuals' autobiographies are often narratives of suffering.

6. See, for instance, Emily Martin's marvelous compilation of Baltimore women's views of their bodies, menstruation, pregnancy, labor, and menopause: Emily Martin, *The Woman in the Body: A Cultural Analysis of Reproduction,* with new intro. (Boston: Beacon Press, 1992).

7. For an excellent example of a modernist conception of disease and science's "battle" against disease "enemies," see Paul deKruif, *Microbe Hunters* (New York: Harcourt Brace Jovanovich, [1926] 1954).

8. Frank, *The Wounded Storyteller,* p. 172.

9. Ibid., p. 173.

10. See, for example, Miles Little, *Humane Medicine* (New York: Cambridge University Press, 1995), pp. 173–175. Little is a surgeon.

11. In this sense, intersexuals' "coming out" reflects the "coming out" of people with other kinds of physical "stigmata," including disabilities; see the "Preface and Acknowledgments," page xvii, in Rosemarie Garland Thomson, ed., *Freakery: Cultural Spectacles of the Extraordinary Body* (New York: New York University Press, 1996).

12. On this history, see especially Ronald Bayer, *Homosexuality and American Psychiatry: The Politics of Diagnosis* (Princeton: Princeton University Press, 1987), and Richard C. Pillard, "The Search for a Genetic Influence on Sexual Orientation," in Vernon A. Rosario, ed., *Science and Homosexualities* (New York: Routledge, 1997), pp. 226–241, esp. pp. 226–229.

13. See, for example, Anne Fausto-Sterling, *Myths of Gender: Biological Theories about Women and Men* (New York: Basic Books, 1992, rev. ed.); Cynthia Eagle Russett, *Sexual Science: The Victorian Construction of Womanhood* (Cambridge: Harvard University Press, 1989); Ruth Bleier, *Science and Gender: A Critique of Biology and Its Theories on Women* (New York and Oxford: Pergamon Press, 1984); Londa Schiebinger, "Skeletons in the Closet: The First Illustrations of the Female Skeleton in Eighteenth-Century Anatomy," *Representations,* 14 (1986): 42–81.

14. Personal correspondence with Diane Marie Anger, 1996–97.

15. Martha Coventry, "Finding the Words," *Chrysalis: The Journal of Transgressive Gender Identities,* 2 (Summer, 1997), forthcoming.

16. Frank, *The Wounded Storyteller,* p. 64. Note also the title of Coventry's autobiographical essay: "Finding the Words."

17. Cheryl Chase, "Affronting Reason," forthcoming.

18. Sven Nicholson, "Take Charge! A Guide to Home Auto-Catheterization," *Chrysalis: The Journal of Transgressive Gender Identities,* 2 (Summer, 1997), forthcoming.

19. Some physicians would argue that Martha Coventry was never "truly" intersexual because her big clitoris alone was insufficient to qualify her as such, but given the similarity of her experiences and those of (other) intersexuals, she sees herself as having been born intersexed; she was certainly treated as having had inappropriate and unacceptable genitalia. Compare Morgan Holmes's declaration, "I'm still intersexual," in *Hermaphrodites with Attitude,* 1 (1994): 5–6. For a discussion of the difficulty in deciding whom to count as "intersexual," see Chapter 1.

20. For summaries and analyses of the present-day protocols, see Suzanne J. Kessler, "The Medical Construction of Gender: Case Management of Intersexed Infants," *Signs,* 16 (1990): 3–26; M. Morgan Holmes, "Medical Politics and Cultural Imperatives: Intersexed Identities beyond Pathology and Erasure" (M.A. thesis, York University, 1994); Anne Fausto-Sterling, "The Five Sexes," *The Sciences,* 33 (1993): 20–25; Deborah Findlay, "Discovering Sex: Medical Science, Feminism, and Intersexuality," *Canadian Review of Sociology and Anthropology,* 32 (1995): 25–52; Anne Fausto-Sterling, "Building Bodies: Biology and the Social Construction of Sexuality" (forthcoming); Chase, "Affronting Reason"; Ellen Hyun-Ju Lee, "Producing Sex: An Interdisciplinary Perspective on Sex Assignment Decisions for Intersexuals" (senior thesis, Brown University, 1994); Anne Fausto-Sterling, "How to Build a Man," in Vernon A. Rosario, ed., *Science and Homosexualities* (New York: Routledge, 1997): 219–225; Patricia K. Donahoe, "The Diagnosis and Treatment of Infants with In-

tersex Abnormalities," *Pediatric Clinics of North America,* 34 (1987): 1333–48; Cynthia H. Meyers-Seifer and Nancy J. Charest, "Diagnosis and Management of Patients with Ambiguous Genitalia," *Seminars in Perinatology,* 16 (1992): 332–339; Arnold G. Coran and Theodore Z. Polley, Jr., "Surgical Management of Ambiguous Genitalia in the Infant and Child," *Journal of Pediatric Surgery,* 26 (1991): 812–820; Patricia K. Donahoe, David M. Powell, and Mary M. Lee, "Clinical Management of Intersex Abnormalities," *Current Problems in Surgery,* 28 (1991): 515–579; Maria I. New, "Female Pseudohermaphroditism," *Seminars in Perinatology,* 16 (1992): 299–318; Kurt Newman, Judson Randolph, and Kathryn Anderson, "The Surgical Management of Infants and Children with Ambiguous Genitalia," *Annals of Surgery,* 215 (1992): 644–653.

21. This concern has been particularly prominent in the work of John Money; see Fausto-Sterling, "How to Build a Man," pp. 222–223; however, the concern shows up in the work of many other intersex experts as well. See, for example, Daniel D. Federman, "Psychosexual Adjustment in Congenital Adrenal Hyperplasia," *New England Journal of Medicine,* 316 (1987): 209–210.

22. See Lee, "Producing Sex," p. 17.

23. For a good summary of Money's approach, see Lee, "Producing Sex," pp. 6–19.

24. Justine Schober, M.D., personal communication, March 5, 1997.

25. See, for example, Barbara C. McGillivray, "The Newborn with Ambiguous Genitalia," *Seminars in Perinatology,* 16 (1991): 365–368, p. 366.

26. Newman, Randolf, and Anderson, "The Surgical Management," p. 651.

27. Ibid., p. 647.

28. Meyers-Seifer and Charest, "Diagnosis and Management," p. 337. See also Stuart R. Kupfer, Charmain A. Quigley, and Frank S. French, "Male Pseudohermaphroditism," *Seminars in Perinatology,* 16 (1992): 319–331, p. 328; McGillivray, "The Newborn," p. 366; Rajkumar Shah, Morton M. Woolley, and Gertrude Costin, "Testicular Feminization: The Androgen Insensitivity Syndrome," *Journal of Pediatric Surgery,* 27 (1992): 757–760, p. 757.

29. See Donahoe, "The Diagnosis and Treatment," p. 1334. Ellen Hyun-Ju Lee has argued that, despite the apparent unity in published medical discourse on intersexuality, "there are no consistent national standards or guidelines for assignment and treatment, only the published work of medical teams at various institutions throughout the country" (Lee, "Producing Sex," p. 37).

30. Robin J. O. Catlin, *Appleton & Lange's Review for the USMLE Step 2* (East Norwalk, Conn.: Appleton & Lange, 1993), p. 49. Quoted with permission. Compare the comments of Dr. Y (an intersex surgeon who wishes to remain anonymous) in Lee, "Producing Sex," p. 24: "[I]t is too hard to make a male."

31. Lee, "Producing Sex," p. 27. Compare this with the story of an intersexed child's surgeries as reported in G. Wayne Miller, *The Work of Human Hands: Hardy Hendren and Surgical Wonder at Children's Hospital* (New York: Random House,

1993). There we read of Beth, a mother who must daily insert dilators into her infant intersexed daughter's

new vagina and rectum [to keep them] from narrowing[.] Beth has to insert a metal dilator deep into each opening once a day, a temporary regimen. Lucy [the daughter] screams the first many times, and her screams unsettle her brother and sister. Beth calls Hardy Hendren [the surgeon] to plead for some other way, a smaller-caliber dilator, anything.

With my luck, the smart aleck in Beth thinks, Lucy will be gay, and all this will have been unnecessary.

"Just do it," Hendren says.

"But—"

"Either you do it, or I have to operate again," Hendren says, and Beth keeps doing it, and eventually the screams stop, as Hendren assured her they would."

(p. 333, with original emphasis)

32. See, for example, M. M. Bailez, John P. Gearhart, Claude Migeon, and John Rock, "Vaginal Reconstruction after Initial Construction of the External Genitalia in Girls with Salt-Wasting Adrenal Hyperplasia," *Journal of Urology,* 148 (1992): 680–684, p. 684. In the discussion reported there, Dr. David Frank asked a panel of intersex surgeons, "How do you define successful intercourse? How many of these girls actually have an orgasm, for example?" Dr. John P. Gearhart, one member of the panel, responded, "Adequate intercourse was defined as successful vaginal penetration." Compare Zoran Krstić et al., "Surgical Treatment of Intersex Disorders," *Journal of Pediatric Surgery,* 30 (1995): 1273–1281: "It is much easier to create a vagina as a passive organ than an erectile phallus with sufficient dimension. Therefore, the authors suggest that most such infants be reared as females" (p. 1273).

33. See Fausto-Sterling, "How to Build a Man," where a surgeon is quoted as remarking, "It's easier to poke a hole than build a pole" (p. 222). Ironically, given the way many intersexuality doctors treat and talk about vaginas, it seems that in their minds only men have vaginas; that is, a vagina is only a true physical entity, an object, an organ when it exists in an intersexed "man" (assigned-male), a situation in which the all-too-present vagina is understood to require removal. By contrast, a vagina need only be a "hole," a constructed space, an *absence* of a thing when the intersexed subject is labeled a woman. See the extraordinary picture of a disembodied vagina—a vagina which had been excised from an intersexual who had been assigned male—included as part of figure 43 on page 64 of Hugh Hampton Young's *Genital Abnormalities, Hermaphroditism, and Related Adrenal Diseases* (Baltimore: Williams and Wilkins Company, 1937); note especially the "photo of specimen removed," labeled drawing 4. In looking at this picture, I realized that I, too, had always envisioned vaginas as holes and not as complicated, responsive organs. This was the case even though I had for years been thinking about Thomas Laqueur's fascinating critique of the very present, phallic-like, winged vagina imagined by the Renaissance anatomist Vesalius. (Laqueur demonstrated, using that example, that our concepts of sex organs are socially constructed; see

Thomas Laqueur, *Making Sex: Body and Gender from the Greeks to Freud* [Cambridge: Harvard University Press, 1990], pp. 78–93; note especially figures 20 and 21.)

34. Kupfer, Quigley, and French, "Male Pseudohermaphroditism," p. 328.

35. Kessler, "The Medical Construction," p. 16; compare the advice given by Meyers-Seifer and Charest, "Diagnosis and Management," p. 336.

36. For a journalist's extraordinary hagiography of an intersex surgeon, a hagiography containing a plethora of restitution narratives, see Miller, *The Work of Human Hands.* For example, on p. 285, Miller writes, "She needed what other little girls had. She needed liberation from this intersex zone to which nature had sentenced her."

37. See, for example, Coran and Polley, "Surgical Management," p. 812; and see p. 363 in P.K. Donahoe and W.H. Hendren III, "Perineal Reconstruction in Ambiguous Genitalia Infants Raised as Females," *Annals of Surgery,* 200 (1984): 363–372.

38. Lee, "Producing Sex," p. 45.

39. See Coran and Polley, "Surgical Management," p. 812; Meyers-Seifer and Charest, "Diagnosis and Management," p. 332; see also the comments of Dr. Y in Lee, "Producing Sex": "[W]e make it a social emergency" (p. 30).

40. Frank, *The Wounded Storyteller,* p. 146.

41. See, for example, the remarks of Antoine Khoury in "Is Early Vaginal Reconstruction Wrong for Some Intersex Girls?" *Urology Times* (February 1997): 10–12.

42. H. Patricia Hynes, "Biotechnology in Agriculture and Reproduction: The Parallels in Public Policy," in H. Patricia Hynes, ed., *Reconstructing Babylon: Essays on Women and Technology* (Bloomington: Indiana University Press, 1991), p. 106. For a similar phenomenon in the medical treatment of conjoined twins, see Alice Domurat Dreger, "The Limits of Individuality: Ritual and Sacrifice in the Lives and Medical Treatment of Conjoined Twins," *Studies in History and Philosophy of Science,* forthcoming.

43. This seems to be particularly true for surgeons. Nonsurgical physicians commonly say that there are three rules of medicine: "If it works, keep doing it; if it doesn't work, stop doing it; and never go to a surgeon unless you want surgery" (Aron Sousa, M.D., pers. comm.).

44. Sydney Segal, "Ethical Issues Presented by Children with Congenital Anomalies," *Seminars in Perinatology,* 16 (1992): 369–373, p. 370, table 1.

45. See Lee, "Producing Sex," pp. 34–35.

46. See Meyers-Seifer and Charest, "Diagnosis and Management": "parents must reach the same decision regarding sex assignment as the experts who have educated them" (p. 338).

47. See Anita Natarajan, "Medical Ethics and Truth Telling in the Case of Androgen Insensitivity Syndrome," *Canadian Medical Association Journal,* 154 (1996): 568–570, pp. 569–570. (For responses to Natarajan's recommendations by AIS women and a partner of an AIS woman, see *Canadian Medical Association Journal,* 154 [1996]: 1829–33.) Compare Shah, Woolley, and Costin, "Testicular Feminization," p. 759.

48. Jan Fichtner et al., "Analysis of Meatal Location in 500 Men: Wide Variation Questions Need for Meatal Advancement in All Pediatric Anterior Hypospadias Cases," *Journal of Urology,* 154 (1995): 833–834, p. 833.

49. Ibid.

50. Ibid., p. 834. Compare Fausto-Sterling's question about what it means to say "premenstrual syndrome" is "abnormal" if, as is alleged by some medical researchers, "70 to 90% of the female population will admit to recurrent premenstrual symptoms" (Fausto-Sterling, *Myths of Gender,* p. 95). Not unlike the findings of Fichtner et al. with regard to hypospadias, in her study of 165 women, Emily Martin "turned up only four women . . . who described themselves as experiencing" PMS (Martin, *The Woman in the Body,* pp. 122–123).

51. Donahoe, Powell, and Lee, "Clinical Management," p. 540.

52. In fact all undescended testicles are at relatively high risk for developing cancer.

53. Letter to the editor by B. Diane Kemp of Ottawa, Ontario, *Canadian Medical Association Journal,* 154 (1996): 1829.

54. H. M. Malin, personal communication of January 17, 1997 to Justine M. Schober, quoted in Justine M. Schober, "Long-Term Outcome of Feminizing Genitoplasty for Intersex," in *Pediatric Surgery and Urology: Long-Term Outcomes,* ed. Pierre D. E. Mouriquand (Philadelphia: William B. Saunders Co., forthcoming).

55. See "In Amerika They Call Us Hermaphrodites," *Chrysalis,* 1997, forthcoming.

56. Anonymous, "Be Open and Honest with Sufferers," *BMJ,* 308 (1994): 1041–42.

57. Quoted from a letter to the editor by Sherri A. Groveman of San Diego, California, *Canadian Medical Association Journal,* 154 (1996): 1829, 1832, see p. 1829.

58. Ibid., p. 1829.

59. The passage in October 1996 of the U.S. federal legislation against female genital mutilation could have an effect on intersex treatment. This law was designed to stop certain African traditionalists from performing "circumcision" on the genitalia of girls under the age of eighteen, whether or not the girls consent. African female genital "cutting" typically involves the mutilation of the clitoris so that most or all of clitoral sensation will be lost, and the stitching together of the labia in order to reduce the size of the vaginal entrance. The latter maneuver aims to increase the pleasure of a woman's male partner, on the belief that a tighter vagina is a more pleasurable one for a penetrating man. Indeed, the process is apparently designed to decrease the woman's sexual pleasure and increase the man's, thereby also reducing the likelihood that a mutilated woman will stray from her partner. (Compare the belief among U.S. intersex doctors that reducing the size of a girl's clitoris will reduce her chances of becoming a lesbian.)

Proponents of this traditional female genital "cutting" have insisted that this practice is an important cultural tradition—a kind of "circumcision" on a girl. But advocates of the U.S. law see it as barbaric and in violation of basic human rights. Representative Patricia Schroeder, who cosponsored the bill along with Senator Harry

Reid, argued that "this is not circumcision. . . . This is more like Lorena Bobbit [who castrated her husband]. Once [other congressional representatives] find out it goes on and is not some victim fantasy we're having, they're horrified" (quoted in Celia W. Dugger, "New Law Bans Genital Cutting in United States," *New York Times,* October 1, 1996, p. A1). Indeed, by 1996 Congress grew horrified enough to declare female genital mutilation unconstitutional and illegal. More particularly, in Section 645 ("Criminalization of Female Genital Mutilation"), Congress declared:

(1) the practice of female genital mutilation is carried out by members of certain cultural and religious groups within the United States; (2) the practice of female genital mutilation often results in the occurrence of physical and psychological health effects that harm the women involved; (3) such mutilation infringes upon the guarantees of rights secured by Federal and State law, both statutory and constitutional; (4) the unique circumstances surrounding the practice of female genital mutilation place it beyond the ability of any single State or local jurisdiction to control. . . . Except as provided in subsection (b), whoever knowingly circumcises, excises, or infibulates the whole or any part of the labia majora or labia minora or clitoris of another person who has not attained the age of 18 years shall be fined under this title or imprisoned not more than 5 years, or both.

Subsection "b" specifies:

A surgical operation is not a violation of this section if the operation is (1) necessary to the health of the person on whom it is performed, and is performed by a person licensed in the place of its performance as a medical practitioner; or (2) performed on a person in labor or who has just given birth and is performed for medical purposes connected with that labor or birth.

Intersex surgeons presumably would argue that the surgeries they perform on the genitals of girls (which clearly include excision of parts of the clitoris) are indeed "necessary to the health of the person on whom it is performed," since intersex surgeons presume small clitorises are necessary to a girl's successful psychosocial adjustment. However, subsection "c" of the law clearly states that "in applying subsection (b) (1), no account shall be taken . . . of any belief on the part of that person [being operated upon], or any other person, that the operation is required as a matter of custom or ritual." We have seen that intersex doctors perform clitoral reduction surgery because they believe it is "a matter of custom" in the United States for girls to have very small clitorises. While it is easy to condemn the African practice of female genital mutilation as barbaric and in violation of the rights of African girls and women, it is perhaps more challenging to ask whether intersex medicine in the United States is also primarily about genital conformity and the "proper" roles of the sexes. Just as we find it necessary to protect the rights and well-being of African girls, we must consider the hard questions of the rights and well-being of children born intersexed in the United States.

60. Newman, Randolf, and Anderson, "The Surgical Management," p. 651.

61. Schober, "Long-Term Outcome."

62. Donahoe, Powell, and Lee, "Clinical Management of Intersex Abnormalities," p. 534. Compare Bailez et al., "Vaginal Reconstruction": "[V]arious procedures have been described for the exteriorization of the vagina and introitus but the long-term results are unclear" (p. 680); see also page 681. See also Tom P. V. M. DeJong and Thomas M. L. Boemers, "Neonatal Management of Female Intersex by Clitorovaginoplasty," *Journal of Urology,* 154 (1995): 830–832: "It will take many more years to prove definitely that the benefits of early operation are maintained throughout puberty" (p. 832). Intersexuals are understandably tired of hearing that "long-term follow-up data are needed" while the surgeries continued to occur. See especially the guest commentary by David Sandberg, "A Call for Clinical Research," *Hermaphrodites with Attitude* (Fall/Winter 1995–96): 8–9; and see the many responses of intersexuals in the same issue and that of Howard Devore, Ph.D., "Endless Calls for 'More Research' as Harmful Interventions Continue," p. 2.

63. See Donahoe, "The Diagnosis and Treatment," pp. 1342–43. Incontinence can also be a problem; see ibid., page 1340.

64. See, for instance, in Donahoe, Powell, and Lee, "Clinical Management": "The dilations that are possible in the young infant can be avoided in the toddler, in whom we have found them impossible to do" (p. 557). This claim is made despite the contrary evidence: "In our experience, this [dilation] maneuver has avoided later stenosis and has given a supple introitus" (p. 554).

65. See Bailez et al., "Vaginal Reconstruction," in which the authors reported that 78.5 percent of their patients required "further vaginal reconstructive surgery to achieve an adequate size to allow for intercourse" (p. 681).

66. "Is Early Vaginal Reconstruction Wrong?" p. 10.

67. Ibid., p. 12.

68. Hormonal treatments come with their own set of problems. Sometimes people born intersexed must have hormonal treatments their entire lives to keep the body type decided by physicians at birth. This results sometimes in psychological distress—over the necessity of medication to "retain" one's identity—and sometimes in dangerous and demoralizing medical "side" effects. It also appears that bodies are limited in their growth spurts, and so inducing growth spurts in young children may nullify any future spurts.

69. See the remarks on this in Coran and Polley, "Surgical Management": "The timing of the surgical procedure represents a balance between the psychological advantages of early surgery and the technical limitations imposed by the small size of the structures" (p. 819). See also Donahoe, Powell, and Lee, "Clinical Management," on vaginoplasty in very young babies: "[In a particular surgical procedure,] one should err on the side of entering the vagina rather than the rectum in order to avoid the catastrophic complication of a rectovaginal fistula. The younger the baby is, the more fragile the vagina and the more difficult the step is" (p. 552).

70. Donahoe as quoted in Lee, "Producing Sex," p. 26.

71. Donahoe, Powell, and Lee, "Clinical Management," p. 544.

72. Justine M. Reilly and C. R. J. Woodhouse, "Small Penis and the Male Sexual Role," *Journal of Urology,* 142 (1989): 569–571, p. 569.

73. Ibid., p. 571.

74. Ibid.

75. Ibid. See also C. R. J. Woodhouse, "The Sexual and Reproductive Consequences of Congenital Genitourinary Anomalies," *Journal of Urology,* 152 (1994): 645–651.

76. Personal communication with Dr. Justine Schober, March 5, 1997.

77. Chase, "Affronting Reason."

78. But sometimes they are killed; on "sacrifice surgeries" in the case of conjoined twins, see Dreger, "The Limits of Individuality."

79. John Money has had a disproportionate influence on the development of intersexuality treatment protocols, but his ideas clearly grew out of his predecessors' (see especially Stephanie Kenen, "'Extraordinary Mix-Up of Sexes May Occur': Hugh Hampton Young and the Surgical Treatment of Hermaphrodites, 1921–1941," in "Scientific Categories of Sexual Difference, 1920–1955" [Ph.D. dissertation, University of California at Berkeley, forthcoming]) and were shaped and bolstered by his contemporaries (see Kessler, "The Medical Construction").

80. The intersexual and transsexual communities are forming many alliances, because they find themselves experiencing inverse problems: transsexuals often cannot get the surgery they want, and intersexuals often have been subjected to surgery they did not want. Nevertheless, transsexualism should not be confused with intersexuality.

81. For nineteenth-century requests, see, for example, the cases of Madame X. and A.H. in Chapter 4, the case of Sophie V. in the Prologue, and the case of Henriette Williams in Lutaud, "De L'Hermaphrodisme au point de vue médico-légal. Nouvelle observation. Henriette Williams," *Journal de médecine de Paris,* (1885): 386–396.

82. Schober, "Long-Term Effects."

83. Quoted in Ibid.

84. Chase, "Affronting Reason."

85. Schober, "Long-Term Effects."

86. Anonymous, "Be Open and Honest."

87. Up-to-date information and referrals are available at the homepage of the Intersex Society of North America (www.isna.org). The Intersex Society of North America (ISNA) may be contacted via P. O. Box 31791, San Francisco, California, 94131, or by electronic mail at "info@isna.org." AEGIS (the American Educational Gender Information Service) is a nonprofit corporation devoted to furthering information about transgender issues and advancing the rights of transgendered persons. Contact AEGIS at P. O. Box 33724, Decatur, Georgia, 30033-0724.

88. "It Is a Folly to Measure the True and False by Our Own Capacity," *Montaigne: Selections from the Essays,* translated and edited by Donald M. Frame (Arlington Heights, Ill.: Harlan Davidson, 1973), lines 36–43.

INDEX

Internal secretions. *See* Endocrinology; Hormones

Intersex, use of term, 4, 31, 39–40, 41, 43, 136–137, 178, 180, 254n19. *See also* Hermaphrodite

Intersex rights movement, 168, 170–173, 176–177, 178, 188, 189, 193, 194, 196, 200–201, 261n85

Kessler, Suzanne, 184, 185, 197

Klebs, Theodor, taxonomy of, *58,* 145–147, 150–151, 155, 247n16,17

Klinefelter's syndrome, 39, 41, 200

Knox, Robert, 144

Krafft-Ebing, Richard von, 133, 134, 138

Labia: varieties of, 35, 36, 37, 38, 39, 54, 60, 85; role in sex diagnosis, 77, 81; problem of diagnosing, 85, 95

Laparotomy. *See* Surgeries, exploratory

Laqueur, Thomas, 209n61, 241n40, 256n33

Laurent, Émile, 119, 138

Lawrence, Thomas. *See* Blacker, George, and Thomas Lawrence

Lee, Ellen Hyun-Ju, 184, 185, 255n29

Lefort, Marie-Madeleine, 54, *55, 56, 57,* 60, 82

Lillie, Frank R., 73, 222n92

Louise-Julia-Anna (case of Guermonprez), 110–113, 114, 131, 138, 187, 239n15

L.S. (case of Tuffier and Lapointe), 130, *131, 132,* 145, 163

Magnan, Valentin, 128–130

Male, various definitions of, 3, 20–21, 27, 34–35, 68, 69–70, 88–89, 102, 106–107, 113–114, 117–118, 122–123, 137, 141, 151–152, 156–157, 164–165, 180, 182–184, 188, 195–196, 199–200. *See also* Pseudohermaphroditism; Sex

Marie-Faustine, 69

Marriages of hermaphrodites: 1–3, 29, *49,* 58–59, 76, 80, 82, 90, 100, 105, 119–120, *121,* 124–126, 128, 163, 175. *See also* Homosexuality, homophobia as impetus for treatments of hermaphroditism

Maternal impressions, 54, 70–71

Mayne, Xavier, 134–135, 136–137, 138

Medical authority, 10–11, 13, 24–25, 35, 45, 60–61, 75, 167–168, 169; challenges to, 10–11, 60–61, 168, 170–173, 178, 180, 188–189. *See also* Medicine, professionalization of; Paternalism; Surgeries, problems with

Medical practitioners: gender of, 10, 24–25, 98; objectification of hermaphrodites by, 46–51; and patient confidentiality, 48–49, 120, 122; relationships with hermaphrodites, 48–50, 52, 57–61, 67–68, 75–78, 99, 107–109, 115–116, 126, 171–172, 175, 176, 178–179. *See also* Medical authority; Medicine, professionalization of; Midwives

Medicine, professionalization of, 10, 11, 52, 60–61, 82–83, 108; rise of hospitals, 64

Menstruation: role of menses and amenorrhea in sex diagnosis, 21, 22, 38, 80, 91, 97–100, 122, 125, 159, 161; in supposed men, 91, 96, 99–100, 108; "compensators" for (molimina), 98–99; "pseudo-menses," 108

Mental deficiency. *See* Insanity and mental deficiency

Micropenis, 191, 195–197, 200

Middlesex Hospital widow (patient of Andrew Clark), 94, 99, 122, 157, 187

Midwives, 25, 60, 82, 215n20

Mistaken sex, 25, 30, 38, 76–78, 79–82, 86, 96–98, 113, 117–118, 197, 206n32; alleged cases of, 1–3, 18, 20–23, *47, 48,* 52–53, 76–77, 80–81, 83, 85–86, 94, 102, 110–113, 117, 118, 120–126, *121, 123, 124,* 128–129, 130, *131, 132,* 206n32, 214n17, 215n20, 230n80, 240n35, 242n62; as explanation for "homosexuality," 76, 126, 129. *See also* Marriages of hermaphrodites; Sex diagnosis

Money, John, 181–182, 261n79

Mothers of hermaphrodites, 58, 71–72, 75, 90, 108, 169. *See also* Maternal impressions

Mullerian system, 34, 69

Neugebauer, Franciszek (Franz): life history, 61–63, *62,* 64; opinions of, *49,* 75, 76, 79–80, 83, 85, 90, 96, 97–98, 117–118, 146, 223n103

Nicholson, Sven, 178–180

Normal and abnormal, problem of defining, 6–7, 28, 41, 115, 177, 184, 188–189, 193, 199, 200, 258n50

Olympic Committee's definition of sex, 7
Orgasm: role in sex diagnosis, 95, 125; treatments' effects on, 174, 192–193, 195, 256n32. *See also* Ejaculation
Ova, 151
Ovaries, 34, 74–75, 126; role in sex diagnosis, 20, 22, 29, 36, 37, 84, 142–145, 152, 155–166; removal of, 66, 93, 98, 99, 152, 217n39, 221n91; embryology of, 69, 151; problem of diagnosing, 85, 86, 91–94, 98–99; cancer of, 128, 129; role in diagnosis of sexuality, 129–130; transplantation of, 159, 224n106. *See also* Endocrinology; Gonadal definition of sex; Hormones
Ovotestes, 36, 37, 73–74, 147, *148,* 155, 159–161, *162*

Paternalism, 187–188, 192
Pelvis, 88, 101–103, 125
Penis: role in sex diagnosis, 1, 20, 22, 77, 80–81, 111, 182, 183, 195–197; varieties, 4–5, 36, 37, 39, 58, 85, 95, 188–189, 193; embryology of, 35; problem of defining, 95–96, 177, 183; surgeries on, 177, 182, 183–184. *See also* Clitoris; Hypospadias; Micropenis; Phallus
Phallus, 4–5, 34, 36, 95–96. *See also* Clitoris; Penis
Physicians. *See* Medical practitioners
Postmodernism, effects on intersex autobiographies, 170–173, 177, 180
Pozzi, Jean Samuel: life history, *63,* 63–65, 189; opinions, 69, 70–71, 75, 94, 98, 101, 105, 108, 114, 128–130, 146, 154–156, 189, 243n69, 248n21, 250n39, 251n52
Prepuce. *See* Penis
Prostate, 91–92
Pseudohermaphroditism: defined, 37, 68–69, 142, 143–145, 155, 223n103; "female" varieties, 37–38, 154; "male" varieties, 38–39, 154; alleged cases of, 54–55, *58,* 83–84, 108–109, 110–113, 114, 128–129,

130, *131, 132,* 173–175; attempts to eliminate term, 155–156; typical sex assignment in cases of, 181–183. *See also* Hermaphrodite; Mistaken sex
Psychological counseling for intersexuals, 180, 190–191, 196, 200–201, 261n85
Psycho-sexual hermaphrodite, use of term, 135
Pubic hair. *See* Hair, pubic

Race, 22, 26, 53–54, 105–106, 136, 211n82
Rémy, Eugénie, *47, 48*

S.B. (patient of William Blair Bell), 159–161, *160, 162,* 163, 166
Schober, Justine Reilly, 195–197, 199, 200
Scrotum, 35, 37, 77, 85; problem of diagnosing, 85, 95
Semen. *See* Ejaculation; Sperm
Sex: as historical ("constructed"), 8, 12–13, 16, 25–30, 42, 43–45, 88–89, 113–114, 127–128, 166, 172–173, 197–198; sexual anatomy, 3–5, 105–106; development of sexual anatomy, 4, 34, 36, 68, 108, 165. *See also* Doubtful sex; Female; Gender; Gonadal definition of sex; Male; Mistaken sex; Sex diagnosis; Sexuality; True sex
Sex assignment: methodology, 37, 81–84, 117–118, 157–159, 161, 163–166, 181–188, 195–197, 199–200, 225n10, 227n36, 239n21, 255n29; cases of reassignment, 17, 18, 52–53, 57–59, 76–77, 80–81, 82, 83, 90, 120–126, 177, 214n17, 240n35; "desexing" procedures, 122–123, 130–131, 157–158. *See also* Gonadal definition of sex; Sex diagnosis
Sex chromosomes. *See* Genetics
Sex diagnosis: general problem of, 13, 23, 28–29, 75–78, 79–80, 81–82, 86; specific techniques used for, 21–22, 60, 67, 82–105, 107–108. *See also* Sex assignment
Sexual inversion. *See* Homosexuality
Sexuality (sexual acts and desires): definition, 10, 88–89, 113, 228n50; role in sex diagnosis, *47,* 88–89, 96–97, 110–117,